D1713644

Sustainable Environments

Sustainable Environments

A Statistical Analysis

Edited by
A.K. Ghosh
J.K. Ghosh
Barun Mukhopadhyay

OXFORD
UNIVERSITY PRESS

YMCA Library Building, Jai Singh Road, New Delhi 110001

Oxford University Press is a department of the University of Oxford It furthers the University's objective of excellence in research, scholarship, and education by publishing worldwide in

Oxford New York

Auckland Bangkok Buenos Aires Cape Town Chennai
Dar es Salaam Delhi Hong Kong Istanbul Karachi Kolkata
Kuala Lumpur Madrid Melbourne Mexico City Mumbai Nairobi
São Paulo Shanghai Taipei Tokyo Toronto

Published in India
By Oxford University Press, New Delhi

First published in India 2003

ISBN 019 565858 2

Printed at Rashtriya Printers, New Delhi 110 032
Published by Manzar Khan, Oxford University Press
YMCA Library Building, Jai Singh Road, New Delhi 110 001

Preface

This volume comprises eight selected contributions on statistical methodology pertaining to environmental issues, including one on global warming and climate change. The contributors are distinguished statisticians or mathematicians working in the field of environment. All these papers were invited presentations to an International Workshop titled 'Statistical Science and Environmental Policy: Possible Interactions', organized by the Indian Statistical Institute (ISI) and the Bernoulli Society for Mathematical Statistics and Probability during 10–12 January 2000, in Kolkata.

The workshop was organized in recognition of the international concern regarding the need to formulate and implement appropriate policies to protect the global as well as local environment. The major environmental theme areas discussed in the workshop were: environmental databases, modelling, and sustainability, global environmental change, pollution, and environment and society. The stress was on statistical models, methods, and innovative computational and inferential tools for design, collection, analysis, and monitoring of environmental data. In the workshop both environmental statisticians as well as environmental scientists participated in order to exchange views and discuss relevant issues from a policy perspective. The papers included in this volume will be useful to researchers and policy-makers in stimulating further discussion and methodological development.

We are thankful to the ISI and the Bernoulli Society for Mathematical Statistics and Probability for organizing the workshop. We are grateful to the numerous individuals and organizations who provided active support for organizing the workshop and in preparing this volume. Special mention must be made of S.B. Rao, K.C. Malhotra, Partha P. Majumder, Robin Mukherjee, S.C. Bagchi, and S. Mukhopadhyay. Each one of them made a unique contribution towards the success of the workshop. We acknowledge the help of our young research scholars in managing the workshop. We are immensely grateful to all the invited speakers, panelists, invited chairpersons, members of our organizing committee and all other young scientists whose participation made the workshop a success. We are extremely grateful to the Ministry of Statistics and Programme Implementation, and the Ministry of Environment and Forests, Government of India, Council of Scientific and Industrial Research, Indian National Science Academy, and West Bengal Pollution Control Board. for financial support. We express our gratitude to K.B. Sinha, Director, ISI, for financial support and encouragement to publish the volume.

We record our appreciation of Dibyendu Bose and Aparesh Chatterjee who helped in preparing the camera ready copy for publication. We are thankful to J. Verghese and his associates in the Reprography and Photography Unit of ISI for their help in converting figures to digitized images. Finally, we are much thankful to the editors of Oxford University Press, India for shouldering the responsibility of publishing this volume.

A.K. GHOSH
J.K. GHOSH
B. MUKHOPADHYAY

Contributors

HARUO ANDO
The Tokyo Metropolitan Research Institute for Environmental Protection, Tokyo, Japan.

ABDEL EL-SHAARAWI
National Water Research Institute, Canada Centre for Inland Waters, Ontario, Canada.

A.K. GHOSH
Centre for Environment and Development, Kolkata, India.

J.K. GHOSH
Indian Statistical Institute, Kolkata, India;
Purdue University, W. Lafayette, USA.

NOBUHISA KASHIWAGI
The Institute of Statistical Mathematics, Tokyo, Japan.

D.R. KOTHAWALE
Indian Institute of Tropical Meteorology, Pune, India.

K. KRISHNA KUMAR
Indian Institute of Tropical Meteorology, Pune, India.

K. RUPA KUMAR
Indian Institute of Tropical Meteorology, Pune, India.

XIAOMING LI
Marck & Company Inc., Blue Bell, USA.

BARUN MUKHOPADHYAY
Indian Statistical Institute, Kolkata, India.

A.K.M. NAZRUL-ISLAM
University of Dhaka, Dhaka, Bangladesh.

KATSUYUKI NINOMIYA
Yokohama Environmental Research Institute, Yokohama, Japan.

BARRY NUSSBAUM
US Environmental Protection Agency, Washington, USA.

HISAKO OGURA
Chiba Environmental Research Institute, Chiba, Japan.

G.P. PATIL
Centre for Statistical Ecology and Environmental Statistics,
The Pennsylvania State University, University Park, USA.

PRANAB K. SEN
University of North Carolina, Chapel Hill, USA.

BIMAL K. SINHA
University of Maryland, Baltimore County, Baltimore, USA.

ROMÁN VIVEROS
McMaster University, Ontario, Canada.

YONGMIN YU
McMaster University, Ontario, Canada.

XIAOGU ZHENG
National Institute of Water and Atmospheric Research,
Wellington, New Zealand.

Contents

List of Abbreviations x
List of Tables xii
List of Figures xiv

1. Introduction
A.K. Ghosh, J.K. Ghosh, Barun Mukhopadhyay 1

2. Upstream-Downstream Water Quality Monitoring
The Niagra River Case Study
Abdel El-shaarawi, Román Viveses, Yongmin Yu 17

3. A Space-time State-space Modelling of Tokyo Bay Pollution
Nobuhisa Kashiwagi, Katsuyuki Ninomya, Haruo Ando, Hisako Ogura 42

4. Statistical Issues in Environmental Evaluations
Bimal K. Sinha, Xiaoming Li, Barry Nussbaum 64

5. Air Pollution
Statistics and Environmental Health Perspectives
Pranab K. Sen 87

6. Mangrove Forest Ecology of Sundarbans
The Study of Change in Water, Soil and Plant Diversity
A.K.M. Nazrul-Islam 126

7. Classified Raster Map Analysis for Sustainable Environment and Development in the 21st Century
G.P. Patil 148

8. Global Warming and Climate Change
An Indian Perspective on Observations and Model Projections
K. Krishna Kumar, K. Rupa Kumar, D.R. Kothawale 175

9. Data Assimilation, an Important and Challenging Research Topic for Environmental Statisticians
Xiaogu Zheng 198

Abbreviations

AISMR	All India Summer Monsoon Rainfall
CEIS	Centre for Environmental Information and Statistics
CFCs	Chlorofluorocarbons
CI	Confidence Interval
CMH	Cochran-Mantel-Haenszel
COA	Canada-Ontario Agreement
COD	Chemical Oxygen Demand
CV	Coefficient of Variation
DJF	December-January-February
DTPSM	Discrete Thin Plate Smoothing Model
3D-Var	Three-dimensional Variational
4D-Var	Four-dimensional Variational
EHS	Environmental Health Science
EKF	Extended Kalman Filtering
ENSO	El Niño Southern Oscillation
EP	Environmental Pollution
EPA	Environmental Protection Agency
EPT	Environmental Pollution and Toxicity
ET	Environmental Toxicity
FACE	Free Air Carbon-dioxide Enrichment
FE	Fort Erie
GAM	Generalized Additive Model
GEE	Generalized Estimating Equation
GHG	Greenhouse Gases
GIS	Geographical Information Systems
GLM	Generalized Linear Model
GPS	Global Positioning System
HMTM	Hierarchical Markov Transition Matrix
ICRP	Indian Climate Research Programme
IGBP	International Geosphere-Biosphere Programme
IPCC	Inter-governmental Panel on Climate Change
IPGW	Iowa Persian Gulf War
ISI	Indian Statistical Institute
JJAS	June-July-August-September
LB	Lower Bound
LRT	Likelihood Ratio Test

MAM	March-April-May
MCMC	Markov Chain Monte Carlo
MLE	Maximum Likelihood Estimate
NID	Normally Distributed Independently of any Other Random Variables
NIEHS	National Institute of Environmental Health Sciences
NIH	National Institutes of Health
NOTL	Niagara-on-the-Lake
NPL	National Priorities List
NRTC	Niagara River Toxics Committee
OEI	Office of Environmental Information
OI	Optimal Interpolation
ON	October-November
PAGES	PAst Global ChangES
PBPK	Physiologically Based Pharmacokinetic
PCB	Polychlorinated biphenyl
PGW	Persian Gulf War
RSPM	Respirable Suspended Particulate Matter
RT	Reproductive Toxicology
RVP	Reid Vapour Pressure
SO	Southern Oscillation
UB	Upper Bound

Tables

2.1	Censoring Patterns for the Ratio of the Concentrations	26
2.2	Results of the Log-normal Regression Fit for PCB in Water at FE	28
2.3	Summary of MLEs for the Log-normal Regression Fits for PCB in Water and Solids at Each Stations	30
2.4	Results of the Log-normal Regression Fit for the Ratio of PCB Concentrations in Water	30
2.5	Results of the Log-normal Regression Fit for the Ratio of PCB Concentration in Solids	31
2.6	Results of Regression-under-transformation Fit for PCB in Water at NOTL	35
2.7	Summary of MLEs for the Regression-under- transformation Fits for PCB in Water and Solids at each Station	36
2.8	Approximate 95 Percent Confidence Intervals for λ for PCB Calculated from the Likelihood Ratio Statistic	37
2.9	Results for the PCB Concentration Ratios in Water	38
2.10	Results for the PCB Concentration Ratios in Solids	38
4.1	Field and Laboratory Data on RVP for Regular Gasoline	76
4.2	Point Estimation and CI for RVP of Regular Gasoline	76
4.3	Simulation Results for Scenarios I and II	78
4.4	Descriptive Measures of the Posterior Distributions of the Parameters	79
4.5	Descriptive Measures of the Posterior Distributions of the Parameters	80
4.6	Summary of Hillsdale Lake Data	84
4.7	Analysis of Results by Logit Model without Covariate	84
4.8	Ninty-five Percent Upper Confidence Limits for Water Clarity	85
6.1	Dominant Vegetation of Three Ecological Zones	135-6
6.2	Analysis of Water Samples from the Bay of Bengal	137
6.3	Physico-Chemical Properties of Soils	138
6.4	Seasonal Variation of Soil Salinity, 1982-1983	138
6.5	Seasonal Variation of Electrical Conductivity in Water Samples of the Rivers of Sundarbans Mangrove Forest	139
6.6	Values of Indices of Plant Diversity for Mangrove and Deciduous Forests	142
6.7	Data of Quadrats in Two Areas of the Oligohaline Zone	143
6.8	Geometric Series Analysis from Data of Quadrats on Four 11m Circular Plots	144
8.1	Summer Monsoon Rainfall Statistics of All-India, Subdivisions and Homogeneous Regions of India during 1871-1990	180
8.2	Occurrence of Droughts and Floods over 29 Meteorological Subdivisions in India: 1871-1994	181

8.3 Areal Extent (Percentage Area of the Country) of Some of the Large-scale (28 percent to 73 percent Area) Droughts and Floods over India during 1871-1994 185
8.4 Linear Trend (^{o}C/100 year) in Mean Temperature and Monsoon Rainfall (mm/years) 191

Figures

2.1	Major Effluent Outfalls and Waste Dumps Adjacent to the Niagara River (from COA 1981)	20
2.2	Plots of the Relative Profile Likelihood (RPL) of λ for PCB	37
3.1	Stereograms of the True Image	51
3.2	Stereograms of the Data	52
3.3	Stereograms of the Trend Estimated from all of the Data	53
3.4	Stereograms of the Trend Estimated from 20 percent of the Data	54
3.5	A Map of Tokyo Bay Area	55
3.6	Time-series Plots of the Estimated Results at Four Monitoring Points, No. 8, 19, 30 and 38	58
3.7	Contour Maps of the Estimated Trend of Salinity in the Surface Layer in Tokyo Bay Every Month from April 1985 to March 1990	59
3.8	Contour Maps of the Estimated Seasonal Components of Salinity in the Surface Layer in Tokyo Bay Every Month from April 1985 to March 1990	60
4.1	Histograms of Estimated Marginal Posterior Distributions of Parameters (a) μ, (b) σ, (c) η, and (d) ρ	79
4.2	Histograms of Estimated Marginal Posterior Distributions of Parameters (a) μ, (b) δ, (c) σ, (d) η and (e) ρ	79
6.1	Agro-ecological Regions	128
6.2	Map of Sundarbans Mangrove Forest showing various Rivers and Locations from where Soil Samples and Water Samples were Collected	134
6.3a	Relative Abundance of Plant Species	141
6.3b	Species Abundance in the Vegetation Study	142
7.1	Schematics of Landscape Fragmentation	154
7.2	Land Cover Maps for Three Watersheds of Pennsylvania	155
7.3	Conditional Entropy Process Profiles as Landscape Fragmentation Profiles for HMTM Models whose Transition Matrices are Obtained from Watersheds with Three Distinctly Different Land Cover Patterns	156
7.4	Conditional Entropy Process Profiles for HMTM Models whose Hypothetical $k \times k$ Transition Matrices have the value λ along the Diagonal and the value $(1-\lambda)/(k-1)$ off the Diagonal	157
7.5	Nested Hierarchy of Pixels	158
7.6	A Comparison between a Single Classification Error Matrix and a Change Detection Error Marix for the Same Vegetation/Land Use Categories	163
7.7	Simulation Result Showing a Possible Two-stage Adaptive Sample	164
7.8	Spatial Patterns of Error for Three Ecosystems	165
7.9	Echelons of Spatial Variation	167
8.1	Network of 306 Raingauge stations and Demarcations of 29 Meterological Subdivisions Considered in the Study	178

8.2 Variation of Monsoon Rainfall Anomaly (percentage of mean) of All India during 1871-1999 184
8.3 Spatial Patterns of Linear Trends in Summer Monsoon Rainfall during 1871-1990 187
8.4 Variation of Annual Surface Air Temperature Anomaly (^{o}C) of All-India during 1991-97 188
8.5 Spatial Patterns of Linear Trends in Mean Annual Surface Air Temperature over India (in ^{o}C per 100 years) during 1901-90 189
8.6 Variation of Mean Annual Temperature (^{o}C) and Summer Monsoon Rainfall (mm) at Major Cities in India 193

1

Introduction

A.K. Ghosh • J.K. Ghosh • Barun Mukhopadhyay

1. An Overview

The environment has been a major concern in North America and Europe for at least thirty years now. During the last decade it has become a major policy issue in developing countries also. While economic development remains a priority, there is an increasing awareness that the process of development, left to itself, can cause irreversible damage to the environment, and that the resultant net addition to wealth and human welfare may very well be negative, if not catastrophic.

With this awareness, the new goal now is sustainable environment-friendly development. This calls for new strategies at various levels - new laws on environmental protection, creation of new databases, creation of professional expertise in various areas of environmental monitoring and evaluation, and generating public awareness, among others. For all this to be possible one needs a quantitative, statistical approach at various levels of sophistication for both data collection and analysis. Consequently, one area where expertise has to be created is that of environmental statistics. Since many of these countries already have well-developed systems of statistical data collection and analysis, all that needs to be done is to provide reorientation and exposure to statistical methodology and data suitable for environmental studies.

Such an opportunity presented itself when the Indian Statistical Institute and the Bernoulli Society organized a workshop on Statistical Science and Environmental Policy: Possible Interactions in

Kolkata in January 2000 as part of the Bernoulli Society's celebrations for the new millenium. The workshop brought together leading international environmental statisticians and environmentalists including environmental economists in India. This volume is a collection of the statistical papers presented at the workshop. They focus on statistical techniques at the cutting edge of the subject of environmental statistics and will be of interest to both specialists and general readers interested in the relevance of statistics to environmental questions. The papers on India's environment and ecology, statistical database, and issues of experimentation will appear in a companion volume, published by the Statistical Publishing Society, on behalf of the Indian Staistical Institute. Together, the two volumes provide a good overview of the environmental issues in developed and developing countries, data needs and databases, and appropriate statistical formulation and analysis of data on complex questions of environmental degradation and improvement and their impact on human beings.

Though protecting the environment means various things to various people, the major issues have been clean water, wetlands, land and forests, air pollution and health, environment in economic decision-making, and, at a somewhat more speculative level, global warming. A few observations may be in order to indicate what one means by global warming and to what extent it is still a matter of speculation. The following material is based on Smith (1993).

The proportion of greenhouse gases, namely carbon dioxide, methane, chlorofluorocarbons (CFCs) and nitrous oxide is increasing in the atmosphere due to industrial growth, deforestation, and other such activities. Since carbon dioxide is the most abundant among these gases, its level is taken as a proxy for the concentration of greenhouse gases in the atmosphere. There has also been a steady increase in the global temperature throughout the last century. These are empirical facts. An element of speculation comes from the general circulation of scientific models which relate high levels of carbon dioxide to the rise in temperature. These models are complex but crude in relation to the complex phenomena that are being modelled. The current predictions, based on these models and a partial adjustment in the light of observational data, suggest a global warming of 1.8^oC by 2030 and 4^oC by 2100. Just as the theory is somewhat speculative because of oversimplifications, the observational evidence of warming is also somewhat ambiguous. For example, it is not easy to separate a trend from a cycle, especially in the presence of long-range dependence in the temperature series.

Wetlands and forests are already well protected by law and public support in the developed countries, though they remain at risk from unscrupulous realtors and natural catastrophes like forest fires. This is not the case in the developing countries, where the problem of conserving wetlands and forests has been further aggravated by continuing growth of population, incomplete land reforms, and inadequate access to resources for the poor. Moreover, the appropriate Acts are either not in place or their implementation is not easy. The importance of biodiversity from the point of view of a sustainable agriculture or intellectual rights and patents is begining to be realized. The Government of India has introduced the Biological Diversity Bill 2000 into Parliament. The Biodiversity Bill 2000 was referred to a Joint Parliamentary Committee which has since approved the bill. It is at present awaiting the approval of the Parliament. Some active environmental groups have moved the courts and forced a passive government to take cognizance of the violations of Acts already in place. Another area that may be of greater interest to the developing countries is sustainable economic development, which imputes, among other things, an economic cost for the use of exhaustible environmental resources that have been traditionally taken as unlimited and free.

The remaining problems of clean air and water and long-run weather forecasts are of common concern to both developed and developing countries. While most of the papers in this volume are in this common area of interest, there is also an innovative article on how to study and monitor wetlands, forests and other land on both long-term and short-term bases. The paper on the ecology of Sundarbans, the world's largest mangrove forest, addresses some of these issues from a classical, ecological perspective.

Some of the major problems in environmental statistics are environmental monitoring, assessment of the impact, for example, of pollution on health or an intervention or Act for improvement, assessing site reclamation, and risk assessment. Most of these problems generate spatio-temporal data, the analysis of which requires a combination of methods for time-series and spatial statistics. There is also the problem of scales. For example, an intervention for reducing risk may be at the level of the central, state, or local government or even a home owner. The data would have such scales too, manifest in different spatial or temporal correlations.

A common model for observed skew distributions of abundance

of a species or concentration of a pollutant is the log-normal distribution. A probabilistic justification of this assumes an independent multiplicative process which, on the log scale, can be represented by a sum of independent random variables. One then invokes the Central Limit Theorem to justify approximate normality in the log scale.

A common model for binary data, for example, one of two possible responses to a question, is the logistic model. It is particularly convenient if one wants to introduce covariates into the model.

It is difficult to make a risk assessment or make recommendations on risk management based on current toxicological data or prospective or retrospective epidemiological studies. Each sort of data has its own problems. Toxicology is a fairly new subject in which one tries to assess the effect of long-term exposure to low-level toxicity, based on animal studies with short exposure to high toxic levels or the effects of accidents, or other such special situations on human health. In epidemiological data, long-term effects on humans are documented, but in retrospective studies there are unknown confounders, and in prospective longitudinal studies, which provide the best data, one has to take into account various forms of censoring. Extreme values and logistic models are frequently used statistical techniques for risk assessment. For example, when one tries to determine a threshold of the magnitude of a flood or an earthquake or any other catastrophic event once in hundred or once in a thousand cases, the models are usually the possible limiting distributions of extreme values. Extreme values can occur more directly if one has several measurements on air pollution and wants to verify if the latest value is unusually high. An alternative way of assessing risk is through logistic models. In this framework, one calculates the probability of a given event based on binary data and also studies the influence of covariates on the risk probability.

An emerging new source of data, available to all, is satellite pictures, which make it relatively easy to study a very large region but require handling very large data sets. The paper by Patil is based on a very large data set but that aspect is not discussed there. A similar application has been made in the Indian context by Gadgil *et al.* (1999, 1998).

2. Clean Water, Air and their Ecological Impact

The first four chapters in this book deal with clean water and air and their ecological impact, especially the impact of the presence or absence of pollution on humans. In Chapter 2, El-Shaarawi evaluates the monitoring programme for the Niagara river in North America, which, in a small stretch of 35 km, connects Lake Erie to Lake Ontario. The famous Niagara Falls are located on this stretch and attract thousands of tourists every year. The river is also ecologically important because it affects not only the towns and villages on its banks but also the vast ecosystem that depends on the waters of Lake Ontario.

The major source of pollution is industry, with products ranging over chemicals, rubber and metal industries, electronics and car plants. In 1984, the governments of Canada, the United States, New York State and the province of Ontario began a detailed water quality monitoring programme 'to assess the effect on pollution levels of policies and by-laws passed by the governments to reduce the amount of pollution dumped into the river'. A sample of water is taken every week at two stations, Fort Erie (FE) and Niagara-on-the-Lake (NOTL), one upstream and the other downstream. The data analyzed by El-Shaarawi covers the period from 1986 to 1996. One is interested in improvements in pollution levels over time at each station and the change in the difference in pollution levels at the two stations over time. Since only two stations are involved, the spatial aspect is not a problem. In principle, there can be some dependence over time so that time-series methods are called for. El-Shaarawi recognizes this, but assumes independence as a first approximation. He focuses on two other difficulties, namely the problem of 'nondetects', that is left censoring, and a proper choice of an error distribution in the regression model of the concentration of a pollutant (PCB) on explanatory variables and time.

A nondetect is declared if the concentration falls below a certain threshold. In this case, all that is recorded is that the threshold was not crossed. This does lead to technical problems. Quite rightly, this is treated as a form of left censoring and is introduced explicitly in the analysis. For error distribution, both the commonly used lognormal for a concentration variable and more flexible methods based on Box-Cox transformations are tried. One reassuring conclusion is that while the latter provide a better fit, the important conclusions based on the log-normal model do not change. El-Shaarawi achieves a delicate balance between the use of sophisticated and novel statistical techniques and communicating the substantial findings relating to practical environmental issues. It is an interesting example of monitoring as well as impact study.

In this volume, Chapter 3 by Kashiwagi *et al.* is about water pollution in relation to the Tokyo Bay. The surface of a bay is a two-dimensional object unlike that of a river which may be thought of as one-dimensional to a first approximation. For a two-dimensional surface, it is natural to approximate by points on a lattice that take into account the boundary of the surface. The monitoring stations would be located there. This seems to be the case if one looks at the monitoring points on the Bay and compares them with the coastal boundary, which has a wall-like effect on the flow of water. This is a natural statistical design, though it is unclear if it is an optimal one. Kashiwagi *et al.* take the design as given and focus on modelling and analyzing the data. One of the main technical objectives was to introduce space and time explicitly into the model, make a seasonal adjustment and then examine for the existence of a trend. Correlation was modelled through an innovative linear model which takes into account the lattice structure with its neighbourhood system as well as time. One complication was the lack of spatial stationarity due to the wall-effect of the coastal boundary on the flow of water. Far from the coast, one would expect spatial stationarity. Missing values complicate things further.

Kashiwagi *et al.* discuss in detail how this complex model can be handled computationally. They also implement a computational scheme for a large data set. This is a major methodological contribution to nonparametric spatio-temporal modelling and seasonal adjustment. Many immediate applications are anticipated.

Chapter 4 by Sinha *et al.* provides a clear formulation of several problems and an interesting analysis with immediate policy implications. One of these problems is concerned with clear, clean water, but not in the usual context of industrial or biological pollution. Hillsday Lake, close to Kansas City, is used as part of a flood control plan for the Osage and Missouri River Basins. The lake is also a major recreational spot; it provides drinking water for two countries. The Environmental Protection Agency of the USA (EPA) conducted a survey of lake users in the summer of 1999 'to establish what level of water clarity users perceive as good'. This will lead to setting up appropriate standards for the total daily sediment load for the lake and its watershed.

The major statistical problem was to set up a regression model relating the proportion of different categories of users' response, like good or excellent, to covariates such as secchi depth. The greater the depth, the more satisfied will a user be. A standard way of doing this is through probit or logistic regression. Logistic regression is the preferred choice of most statisticians; Sinha *et al.* appear to have used both models but present only the results based on the logistic model.

One application of this analysis is to estimate the depth which would lead to a given proportion, say 80 per cent, of satisfied users (users who perceive the condition of the lake as good or better). This would provide the sort of concrete environmental guideline that the EPA can use to set up its standards for the secchi depth of the lake and the watershed.

In another paper on the same problem, Sinha *et al.* (2001) explore a double exponential rather than a logistic link function. The analysis shows that the double exponential model works best, followed by the logistic and the probit, in that order. They also obseve that the introduction of temperature as an additional covariate improves the model. The lesson seems to be, as in Chapter 2, that while traditional models like logistic or log-normal are adequate for practical purposes, one can get a better fit by exploring beyond them. The extra fit can matter in detecting small changes.

The other two problems considered in Chapter 4 do not deal with water quality but provide novel applications of statistical techniques. One such application is to estimate a common mean of two sets of data with different amounts of variability. We deal with this application first. The statistical model involves a pair of random variables (y, z) with same mean and same dispersion matrix. The data consists of m pairs of observations on $(y < z)$ along with additional data on one of the variables, say y. It is the latter feature which requires new work.

The model is applied to the EPA's evaluation of gasolene quality, based on Reid Vapour Pressure (RVP). This is part of the programme to reduce air pollution by use of gasolene with reduced volatility. To ensure compliance, samples of gasolene are taken from various pumps and measured in two ways–at the field level, in a relatively cheap but possibly inaccurate way, and at the laboratory level. One may think of these as y (measurement on field) and z (measurement in laboratory). There are also some field-only data (y).

Sinha *et al.* provide various methods of estimating the common mean as well as the variance of y and z. Their analysis also reveals an unexpected finding of practical importance, namely that the estimates based on field–only data are likely to be downward biased, in addition to field data having more variability. The first fact suggests that the EPA should continue to make random spot checks in the laboratory and keep in mind the possibility of bias in field measurement when setting up their standard for RVP.

Another application addresses the issue of evaluating a reclaimed site that was earlier used for hazardous waste storage. The basic statistical problem is somewhat similar to that in El-Shaarawi, in that quality measurements are to be compared at two points of time, one of which is before or at the begining of a major intervention and the other is later. This problem is about the possible dependence between the two sets of measurements and the impact of dependence on modelling and analysis. An interesting new model is proposed. This issue needs to be studied further both empirically and statistically.

Chapter 5 by Sen, the last in this set of four chapters on water and air pollution, is a definitive overview of a broad range of statistical problems that need to be solved before we can make a reliable risk assessment of the toxicological impacts of air pollution. Sen identifies three main ways air pollution causes health hazards, namely inhalation, ingestion and absorption. He also identifies three components of quantitative risk assessment, namely (i) quantification of air pollution level, (ii) dosimetric studies, usually on animals, to acquire scientific evidence, and (iii) observational studies to relate exposure to pollution and health problems. Sen suggests an integrated approach, which combines (ii) with (i) and (iii), would be far more promising than the current tendency to use (ii) for simple-minded extrapolation to humans, and use existing statistical methods separately for each area. Risk assessment in this context is seen to be a complex scientific problem. An integrated scientific approach will strengthen conventional statistical analysis as well as scientific models by adding expert information and judgement. Data-based statistical models, for example those based on regression, are parsimonious and have good predictive power for similar data. But their explanatory power is less, and hence extrapolation to new ranges of covariates or different experiments is problematic. On the other hand, scientific models have good explanatory power but tend to be over-parametrized, leading to inadequate prediction. Sen would like to see an integration of the two approaches to arrive at a model which preserves the salient features of a scientific model but simplifies the parametrization. A natural way of implementing this is through Bayesian analysis, but Sen seems to be worried about possible oversimplification and the consequent lack of robustness in the statistical analysis. It is of course too early to make a firm statement on this.

In accordance with Sen's emphasis on an integrated approach, Chapter 5 discusses at length how air pollutants reach and damage human organs, in this case mainly the respiratory and cardiovascular systems and the skin. The carcinogenic and other impacts of pollutions, where partial scientific understanding is available, are also discussed. The chapter provides an overview of the current problems and possible statistical solutions for the assessment of toxicity and risk associated with environmental pollution. Sen makes a number of interesting suggestions, which are new but have been independently used by the other contributions to this volume as well. For example, in spatial studies he suggests choosing a grid or lattice of points that reflects the geometry of the polluting process and the lack of stationarity-both these seem to have been taken care of well by Kashiwagi *et al.* in Chapter 3. Sen's concern for the lack of normality and appropriate use of Box-Cox transformation finds justification in El-Shaarawi's able handling of these issues in Chapter 2. However, Sen's concern about non-normality of count or concentration data goes further. He feels even transformations or generalized linear models (for example, the logistic linear model in Chapter 4) may not be adequate, and suggests exploring generalized additive models in this context.

Sen then focuses on the difference between monitoring studies, as illustrated by Chapters 2–4, and the relatively recent studies on the effect of air pollution on human health. He discusses in detail the possibility of confounders, and the consequent need for more care in data collection in studies on environmental pollution and toxicity (EPT) than is common in monitoring studies. An additional problem is that the short-term effect on health, which is used as a surrogate for the long-term effect, is usually negligible. It is worth mentioning in this context that a quite successful investigation of this kind was presented at the workshop, but is reported elsewhere (Hwang *et al.* 2000). That study relates changes in the weather and levels of air pollution to the illness of school children and their absence from school at those points of time. Possible confounders are also taken care of. This delicate, careful and innovative analysis reveals a significant effect of air pollution on health something not demonstrated so well before. The analysis vindicates many of Sen's concerns.

Sen provides an illustration of the integrated approach to risk assessment and also lists some of the formidable challenges to satisfactory statistical analysis - non-linear regression, lack of well-tested models, need for robustness, lack of information on covariates or exposure to risk at the level of individuals under study. A partial remedy for the lack of information on exposure at an individual level is the newly available pollution maps. Sen also discusses the importance of using biomechanistic and physiologically based pharmacokinetic (PBPK) models in clarifying environmental toxicologial studies and bridging the gap between dosimetric studies on animals and clinical studies on humans.[1] A method that needs special attention is the application of Bayesian ideas, especially hierarchical Bayesian and MCMC (Markov Chain Monte Carlo), to provide realistic formulations of complex environmental problems. A good example of what can be done in the rather different context of radon emission is the study by Gelman *et al.* (1999).

Sen provides interesting comments on existing methods as well as new thoughts, plus a substantial bibliography.

3. Ecological Degradation and Biodiversity

Chapters 6 and 7 address the issues of ecological degradation and biodiversity. For readers in developing countries, these are among the most vital issues in environmental studies. The tropical developing countries account for most of the biodiversity of our planet, but it is also there that many species are rapidly becoming extinct. Linked with this concern are apprehensions about the impact of the agricultural technology of the Green Revolution, based on chemical fertilizers, high-yielding varieties, and pesticides. The negative effect on soil quality and biodiversity is beginning to be seen as a very high price for the undoubted benefit of food security ushered in by the Green Revolution.

The Sundarbans form an extensive forest-covered region in the southern part of the delta of the river Ganges, falling partly in Bangladesh and partly in the state of West Bengal in India. Both

[1] The reader would benefit from reading, along with this review, the illuminating papers on toxicology in the 1988 volume of *Statistical Sciences*.

parts are biological hotspots - they are well- known for their mangrove trees, nesting Olive Ridleys, and Royal Bengal Tigers. Most of the water is saline owing to the region's proximity to the Bay of Bengal, and the trees and animals in the region have adapted to this. While both parts, that is those in Bangladesh and West Bengal, show degradation over time, the degradation in West Bengal is greater. A major cause of this degradation is the increased salinity of water due to a reduced volume of water in the these regions.

Chapter 6 by Nazrul-Islam describes various aspects of the Sundarbans and examines the biodiversity of plants in the region through a number of well-known indices for diversity. The chapter contains a wealth of data. Nazrul-Islam will hopefully return to this problem and explore the spatio-temporal variation of species, abundance and biomass and relate them to variation in resources. Some interdisciplinary work of this kind has been done by a group at the Indian Statistical Institute, Calcutta for the Sunderbans in West Bengal. With the help of statisticians, Manoranjan Ghosh and his students have studied spatial variation and related this to salinity and dryness. There is more salinity away from the coast because the discharge of the rivers does not reach these regions.

Chapter 7 by Patil is primarily based on remote sensing data on three watersheds in the state of Pennsylvania in USA. The object is to use new statistical methodology as well as information technology to provide easily seen and understood summaries to answer many questions on spatial pattern and scales as well as the current status of land and trend of evolution, degradation and fragmentation. The intended readers are not just environmental scientists and statisticians but town planners, forest managers and other policy-makers engaged in similar activities. Like all ambitious work, it will need careful exploration, validation, and evaluation by readers and future users before it can be recommended as a ready-made tool. But at this time, it certainly appears to be an exciting, promising new approach.

The basic approach is as follows. The chosen landscape is represented by a number of land cover types at different scales, which can be thought of as states of Markov transition models at different scales. One seeks a stochastic model which is simple but provides a realistic description of the changes in the landscape fragmentation. This measure is in turn an index of degradation and a proxy for risk calculation.

The stochastic model is slightly more complex than just a Markov chain for it provides for both fragmentation as well as changing pattern of land cover. This is done as follows. A pixel at a particular scale gives rise to four pixels at the next finer scale. The transition probability of the latter to belong to a particular land cover depends only on the land cover type of the former. The land covers for the four smaller pixels are determined by four independent draws. The data-based choice of the transition and splitting probabilities is also novel and sophisticated. This sophisticated yet simple model and its parameters, along with their physical interpretation, provide the promised summary. It would be interesting to apply and validate these methods on the land-cover data on the Western Ghats given in Nagendra and Gadgil (1999, 1998).

In their application of the above methods to the Western Ghats, Gadgil and his group converted the basic satellite data on pixels into a vegetation index. The pixels were combined to form a superpixel, the scale of the latter being somewhat arbitrary. For each superpixel, they calculated the frequency distribution of the vegetation indices of the constituent pixels. Four summary measures, namely, mean, standard deviation, and coefficients of skewness and kurtosis were taken as the variables for each superpixel. Using cluster analysis, some expert judgement, and comparison with a certain amount of field data, they prepared a map for the region with fourteen types of land covers. Gadgil and his group expect that such maps will be prepared from time to time to monitor the kinds of changes taking place.

A deep study of this map can be made along the lines indicated by Patil. One could examine what physical interpretations can be attached to the model, its parameters including the eigen values of the transition matrix and indices of biodiversity. Such an analysis would provide insight about the data on Western Ghats as well as the methods being applied.

Another way of validation would be to estimate scales at different levels in a fairly direct way, using variograms or correlograms. One could then compare the scale estimates with those obtained by the method of Patil. Another possible approach to estimating the scales is the application of point processes.

4. Climate and Weather Changes

Climate and weather changes are part of our changing environment. Chapter 8, by Krishna Kumar *et al.* discusses in detail global warming and climate change in India, while Chapter 9 by Zheng suggests how modelling and prediction problems can be better resolved through data assimilation.

Chapter 8 identifies a few variables like rainfall and maximum and minimum daily temperature and examines their temporal and spatial variation. Rain in India comes mostly during the monsoon, an event of great importance in the Indian economy, life, and culture. The long series of data, going back to more than hundred years, does not indicate any trend or lagged correlation. Though the coefficient of variation of total rainfall is small, there is a great deal of spatial variation. An interesting fact is the high correlation between years of unusual flooding or droughts and events like the El Nino and La Nina coupled with the Southern Oscillation.

The series on daily temperatures shows a steady warming trend in the all-India mean annual temperature. A similar trend is also visible in the series on the maximum temperature but much less so in the series on the minimum temperature. This phenomenon is especially interesting in the context of a global trend of diminishing difference between maximum and minimum temperatures. Chapter 8 also presents some simulations of the future, based on long-term deterministic models with no stochastic component. It provides a valuable overview of Indian meteorological data.

In Chapter 9 Zheng urges close collaboration between statisticians and meteorologists in modelling and predicting data. He shows how statistical ideas essentially based on solving high-dimensional partial differential equations have improved weather prediction. Zheng observes that the dynamical system being modelled is chaotic. Hence the solution is sensitive to even slight changes in a deterministically fixed initial condition. A better strategy is to look at the relevant ensemble of initial conditions based on observed data. The average over this ensemble seems a better starting point and has been seen to work well.

A second problem relates to handling a high-dimensional evolution matrix that emerges from the dynamical system. In the 1980s, this was solved by defining neighbours and taking averages (as in Chapter 3, for example). However, it leads to unacceptable magnitudes of errors in highly non-linear systems. Zheng discusses how variational methods developed in the 1990s may provide better approximations. Data assimilation apears to be a name for these innovative ways of handing non-linear, high-dimensional, chaotic dynamic systems of meteorology. Zheng's methods are at the cutting edge of the subject.

Another interesting study by Filar shows the relevance of introducing a stochastic error term in the deterministic European general circulation models regarding global warming. Filar shows that the presence of such realistic terms can alter the predictions of the model substantially. While this should not mean that one should take the prospect of global warming lightly or back out of national and international commitments to reduce the emission of CO_2, it is good to be aware of the limitations of the current data and modelling and prediction (Filar and Zapert 1996). Similar introduction of stochastic elements, including non-informative prior distributions, by Mark Berliner and others, have led to preceptible improvements in prediction as well as probabilistic error limits.

One expected effect of global warming is to raise the sea levels. Both in India and elsewhere, estimates are now available on how much coastal land will go under water. Another interesting set of new ideas and experiments worth noting are those in the context of the effects of global warming. Free air carbon dioxide enrichment (FACE) is provided into natural ecosystems to study the effect of an enhanced supply of carbon dioxide.

From this book, the reader will get a fairly good idea of the major areas of environmental statistics, emerging new methodology, and an overview of the future. Blending sophisticated statistical analysis with clearly formulated major environmental problems, the analysis has substantial policy implications for cases such as air and water pollution, monitoring biodiversity and global warming.

This book provides new methods that would lead to further research and insights into many practical problems. Consider the question of the optimal location of monitoring stations. A good statistical formulation covering many contexts would be a major contribution, something that was also discussed at the workshop. The models for data used in this volume would play an important role in formulating these design problems. Many other practical problems of this kind that can be handled by the methods of this book will occur to this reader. Reading the book could be the beginning of such a dialogue between an environmentally aware reader and the distinguished statisticians whose articles appear in this volume.

REFERENCES

Filar, J.A. and R. Zapert (1996), 'Uncertainty Analysis of a Greenhouse Effect Model', in C. Carraro and A. Haurie (eds), *Operations Research and*

Environmental Management, Dordrecht: Kluwer Academic Press.

Hwang, J.S., Y.J. Chen, J.D. Wang, Y.M. Lai, C.Y. Yang and C.C. Chan (2000), 'A Subject-Domain Approach to Study of Air Pollution Effects on School Children's Illness Absence', *American Journal of Epidemiology*, 152, pp. 67-74.

Lin, C., A. Gelman, P.N. Prince and D.H. Krantz (1999), 'Analysis of Local Decisions using Hierarchical Modeling, Applied to Home Radon Measurement and Remediation', *Statistical Science*, 14, pp. 305-33.

Nagendra, H. and M. Gadgil (1999), 'Biodiversity Assessment at Multiple Scales: Linking Remotely Sensed Data with Field Information', *Proceedings of the National Academy of Sciences*, 96, pp. 9154–8.

Nagendra, H. and M. Gadgil (1998), 'Linking Regional and Landscape Scales for Assessing Biodiversity: A Case Study from Western Ghat', *Current Science*, 75, pp. 264-71.

Sinha, B.K., X. Li and B. Nussbaum (2001), 'A Statistical Analysis of Hillsdale Lake Data', *Calcutta Statistical Association Bulletin*, 50, pp. 293-306.

Smith, R.L. (1993), 'Long-range Dependence and Global Warming', in V. Barnett and K.F. Turkman (eds), *Statistics for the Environment*, Chichstern: Wiley.

2

Upstream-Downstream Water Quality Monitoring
The Niagara River Case Study

ABDEL EL-SHAARAWI • ROMÁN VIVEROS • YONGMIN YU

A typical approach for evaluating temporal and spatial changes in the water quality of a river is to monitor the concentrations of relevant pollutants routinely at a sequence of hierarchical locations. Thus, an upstream location serves as a reference for a downstream location indicating how water quality has been modified by input sources between the two locations. This paper uses the Niagara River monitoring programme as a template to present a general framework for the statistical analysis of data generated by this approach. We begin with a brief history of the river's pollution problem, the monitoring programme and its objectives. Difficulties regarding data analysis are emphasized and addressed. A likelihood-based methodology is presented for analysis of the data on pollutant concentration and their ratio. Polychlorinated biphenyl (PCB) data from the upstream and downstream locations of the Niagara River are used to illustrate the method. Areas where more work is needed are indicated.

1. INTRODUCTION

Water pollution, particularly by toxic contaminants, is a major threat to the fabric of life and human health. To identify the problem, its extent, and to measure progress towards its elimination requires the establishment of a monitoring system which is capable of producing

---

This research was supported in part by a grant from the Natural Sciences and Engineering Research Council of Canada.

relevant, accurate and timely spatial and temporal measurements on pollution levels. Because a river's flow is frequently unidirectional, upstream/downstream monitoring of pollution is the most common approach. For example, the Niagara River Toxics Committee (NRTC 1984) adopted this approach for the long-term monitoring of the river. We shall use the Niagara river as a model for discussing the issues involved in the analysis of toxic contaminants' data and specifically to estimate quantities of interest to NRTC.

The paper is structured as follows. Following the introduction, Section 2 reviews the river's relevant characteristics, its pollution sources and the associated concerns. Section 3 provides information on the Upstream/Downstream Monitoring Programme of the Niagara River and its objectives. The important issue of the presence of nondetects in the data due to limitations of the measurement process is discussed in Section 4. Models based on the widely used log-normal distribution are presented in Section 5 for left-censored data in the presence of explanatory variables. Furthermore, the Box-Cox transformation is also introduced to expand the range of the model's applicability to the cases when the log-normal model is inadequate. Parameter estimation and inferences are discussed in Sections 6–7 using likelihood-based methods. The direct modelling of the pollutant concentration at a single location and the modelling of the ratio of the pollutant concentration at two matched locations are the two main cases considered in Section 6. Applications to the Niagara River's polychlorinated biphenyl (PCB) concentrations in water and solids from the upstream and downstream locations are presented and discussed. Proposals for future research and extension are given in Section 8.

2. The Niagara River

A stretch of 35 km of running water that connects Lake Erie to Lake Ontario, the Niagara River serves as a short section of the vast border between USA and Canada. The river's in-flow point is in the town of Fort Erie, its water is discharged into Lake Ontario at the town of Niagara-on-the-Lake. Near the middle of the river are the Niagara Falls, one of the world's most famous and spectacular natural wonders, visited by more tourists than any other site in the Great Lakes region.

Unfortunately, the Niagara River has endured extensive dumping of chemical waste. This adds to the contaminants already present in the water from chemical dump sites on the shores of Lake Erie. The sources comprise a large number of industrial sites ranging from chemical companies, pharmaceuticals, rubber and heavy metal industries to electronics and car plants. Vincent and Franzen (1982) compiled a detailed list of the names and types of industrial sources of chemical substances in the Niagara River area, their location and ranking regarding the level of contamination they generate. Allan *et al.* (1983), who provide a concise and informative introduction to the pollution problems in the Niagara River, note that 15 of the major industrial sources collectively discharge 95 percent of the load of priority pollutants released directly into the Niagara River. The remaining 5 percent of the direct discharges comes from smaller diffuse sources. Canadian direct discharge sources of organic priority (serious and needing imediate attention) pollutants contribute less than 3 percent of the U.S. total direct inputs. Heavy metals contributed directly from Canadian industrial sources are about 13 percent of the total direct U.S. load.

The risks and dangerous effects of the contaminants in the Niagara River resonate not only in the river's vicinity but also in the huge ecological and human populations that live along or depend on the waters of Lake Ontario. This is due to the paramount impact that the Niagara River has on Lake Ontario. For instance, twenty-six billion m^3 of water and 4.8 million *tons* of fine grain sediment enter the lake each year from the river (Kaminsky *et al.* 1983). This accounts for 83 percent of the total water (Eadie and Roberton 1976) and 50 percent of the lake's total sediment input (Kemp and Harper 1976). Frank *et al.* (1979), Haile (1977) and Holdrinet *et al.* (1978) who have studied the contamination of Lake Ontario by the Niagara River, have identified adsorption of suspended solids as the main mechanism by which heavy metals and pesticides are spread and ultimately accumulated. The situation prompted major actions undertaken by the province or state and federal governments involved.

3. Water Quality Monitoring Programme

In 1984, the governments of Canada, USA, New York State and the Province of Ontario adopted a water quality monitoring programme with regard to the variety of contaminants that reach the Niagara River from sources in the river's vicinity. The programme is funded and operated by Environment Canada. Sixty-seven chemicals were targeted for monitoring. The actual monitoring continues upto the present day with an expanded list of at least 100 chemicals. The programme began full operation in April 1986.

A sample of water is taken weekly at each of two stations, one located at the head of the river in Fort Erie (FE) and the other one at the river's mouth in Niagara-on-the-Lake (NOTL). The suspended solids are separated from the water in each sample. The concentration of solids and of each chemical in the water and in the solids are measured. Water flow is also recorded. The samples at the two stations are timed about 16-18 hours apart, roughly the water travelling time from FE to NOTL. For the most part, the sample collection at FE is done on Tuesday afternoon of each week. The map in Figure 2.1 displays the geographical location of the major effluent outfalls and waste dumps adjacent to the Niagara River, as well as the location of the two monitoring stations.

The original main objectives of the Water Quality Monitoring Programme were:

(1) To provide an accurate quantitative assessment of the water pollution levels in the Niagara River each year.

(2) To assess the effect on the pollution levels of policies and by-laws passed by the governments involved to reduce the amount of pollutants dumped into the river.

(3) To make objective assessments on whether the pollution levels fell to the specified target values set for subsequent points in time. One of such targets was to achieve a reduction of 50 percent in the levels of selected dangerous chemicals 10 years down the road, that is, by 1996.

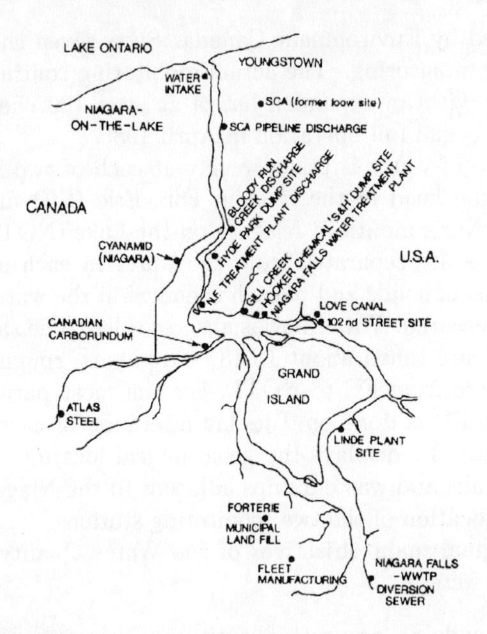

Figure 2.1: Major Effluent Outfalls and Waste Dumps Adjacent to the Niagara River

Source : COA 1981.

4. The Presence of Nondetects

The measured pollutant concentrations in water samples from the Niagara River contain a large number of *nondetects*. These refer to situations where the chemist is unable to quantify the concentration in a sample, usually because of its low level. Small concentrations of a highly toxic pollutant may still pose a risk to the public. Care must be exercised in accounting for nondetects in any statistical analyses of the data.

In making a decision, the chemist uses a *detection limit* for every pollutant. When a measured concentration falls below the detection limit, a nondetect is reported. Refer to the Niagara River Analytical Protocol (1987) for a detailed discussion of the detection limits used in the Niagara River pollutant concentrations. They are known as *practical detection limits* and are determined as follows. Five samples of unpolluted water, that is, samples without the chemical, are prepared. The concentration of the chemical, measured by the method and instrument used in the data collection, is then determined for each sample. The sample average $\bar{x}$ and standard deviation s are then calculated. The detection limit d for that chemical is set as the 95 percent upper confidence bound for the mean concentration of the chemical in unpolluted samples, that is,

$$d = \bar{x} + t_{0.05,4}s = \bar{x} + 2.132s \tag{2.1}$$

Updates on the detection limit for each chemical were done at least once during the 10-year period (2 April 1986–27 March 1996) of data analyzed in this article.

Detection limits are intended to describe how well an analytical protocol discriminates low signals of a pollutant from background noise. They are considered in situations where the instrumentation registers a low signal, but the chemist decides that unpolluted samples could give comparable signal concentrations. However, there is no general agreement on how to handle them in statistical analyses of data (Currie 1988). A serious statistical discussion of the possibilities is given by Lambert *et al.* (1991).

Another feature of the Niagara River pollutant data is the presence of missing values. The general pattern is the absence of all the data for certain samples. The extent of absence is much smaller than that of nondetects.

5. Statistical Models and Handling of Nondetects

5.1 Log-normal Regression

Let Y denote the concentration of a given contaminant in a sample of water, as measured in the original scale, and $x_1, x_2, \ldots, x_p$ be explanatory variables measured along with Y. The log-normal regression model for Y on $x_1, x_2, \ldots, x_p$ can be written thus

$$lnY = \beta_0 + \beta_1 x_1 + \ldots + \beta_p x_p + \epsilon, \epsilon \sim N(0, \sigma^2)$$

In other words,

$$Y \sim \textit{Log-normal}\ (\mu_x = x'\beta, \sigma^2)$$

where $x = (1, x_1, x_2, \ldots, x_p)'$ and $\beta = (\beta_0, \beta_1, \ldots, \beta_p)'$. The regression parameters β and the variance per log observation σ^2 are all treated as unknown parameters. Thus, the probability density function for Y is

$$f(y;\sigma,\beta|x) = \frac{1}{\sqrt{2\pi}\sigma y} e^{-(lny-X'\beta)^2/(2\sigma^2)}, 0 < y < \infty \quad (2.2)$$

As for the distribution function of Y,

$$F(y;\sigma,\beta|x) = Pr(Y \leq y) = \Phi\left(\frac{lny - x'\beta}{\sigma}\right), \quad (2.3)$$

where $\Phi(z)$ is the standard normal distribution function,

$$\Phi(z) = \int_{-\infty}^{z} \frac{1}{\sqrt{2\pi}} e^{-u^2/2} du, -\infty < z < \infty$$

Note that for the concentration in the original scale, the mean and variance are

$$\mu_Y = e^{x'\beta+\sigma^2/2}, \text{ and } \sigma_Y^2 = e^{2x'\beta+\sigma^2}(e^{\sigma^2} - 1)$$

5.2 Handling of Nondetects

In the earlier years of operation of the Niagara River Water Quality Monitoring Programme, the issue of how best to account for the nondetects received considerable attention. Replacing a nondetect by $0, d/2$ or d, where d is the detection limit, were the alternatives that received the widest support. El-Shaarawi (1989) examined these alternatives and concluded, with ample numerical and theoretical evidence, that treating nondetects as left-censored observations was the most sensible course of action. Thus, the only objective information for the data analyst from a nondetect is that $Y \leq d$. Maximum likelihood estimation (MLE) incorporates censored observations in a natural manner (Lawless 1982, El-Shaarawi 1989). Note that Lambert *et al.* (1991) consider left-censoring a credible way to handle nondetects, among others. We adopt the censoring approach in this article.

5.3 Beyond Log-normality: Regression Under Transformation

The log-normal distribution for pollutant concentration data finds its justification in the fact that the concentration of a pollutant in the environment is the result of many slight dilutions (see, for example, Ott 1995, 1990). The multiplicative effect of dilutions becomes additive when transformed to the logarithmic scale. The central limit theorem can be used to justify the normal approximation for the distribution of the log-concentrations. This, however, is only an approximation. Schmoyer *et al.* (1996) cast some doubts on the validity of the log-normal model and provide many 1-sample data sets on pollutant concentrations for which the log-normal model is too heavily right-tailed for a satisfactory description.

In this article, we allow for departures from log-normality by considering an adaptive family of data transformations leading to normality. This family includes the logarithmic transformation as a special case. Specifically, the basic assumption is that there is a value of λ $(-\infty < \lambda < \infty)$ such that

$$\frac{Y^\lambda - 1}{\lambda} \sim N(\mu_x = x'\beta, \sigma^2)$$

This is the well-known Box-Cox family of transformations (Box and Cox 1964), where $\lambda = 0$ corresponds to the log-normal regression model. We treat λ as an unknown parameter requiring estimation. Within the larger family, one can assess the adequacy of the log-normal model, a task that is difficult to accomplish by standard methods (e.g. Q–Q plots) when censored observations are present. Shumway *et al.* (1989) proposed and illustrated this approach with several 1-sample pollutant concentration data with non-detects. Thus, our methods here extend Shumway *et al.* (1989) to the regression context.

According to the above assumption, Y has the probability density function

$$f(y; \lambda, \beta, \sigma \mid x) = \frac{y^{\lambda-1}}{\sqrt{2\pi}\sigma} exp\left[-\frac{((y^\lambda - 1)/\lambda - x'\beta)^2}{2\sigma^2}\right] \tag{2.4}$$

and the cumulative distribution function

$$\begin{aligned} F(y; \lambda, \beta, \sigma \mid x) &= Pr(Y \leq y) \\ &= \Phi\left(\frac{(y^\lambda - 1)\lambda - x'\beta}{\sigma}\right) \end{aligned} \tag{2.5}$$

6. Log-normal Regression for Censored Data

6.1 Single-Station Analysis: Log-Likelihood and ML Equations

Let y_i denote the observed measured concentration of the pollutant on the ith sampling date for one sample of water or solids at one station, and $x_i = (1, x_{1i}, X_{2i}, \ldots, x_{pi})'$ the corresponding $(p+1) \times 1$ vector of explanatory variables. Let d_i denote the pollutant's detection limit on the ith occasion. We assume that the measured concentrations are independent. Following Lawless (1982), p.36, the joint likelihood of

$\theta = (\sigma, \beta_0, \beta_1, \ldots, \beta_p)'$ can be written as

$$L = L(\theta) \propto \left[\prod_{i\epsilon A} f(y_i; \theta \mid x_i)\right]\left[\prod_{i\epsilon B} F(d_i; \theta \mid x_i)\right] \tag{2.6}$$

where A is the set of dates on which observations above the detection limit were made and B those dates resulting in nondetects. Replacing (2.2)–(2.3) in (2.6) yields the joint log-likelihood

$$\begin{aligned} lnL &= -n_A ln\sigma - \frac{1}{2}\sum_{i \in A}\left(\frac{lny_i - x_i'\beta}{\sigma}\right)^2 \\ &\quad + \sum_{i \in B} ln\Phi\left(\frac{lnd_i - x_i'\beta}{\sigma}\right) \end{aligned}$$

where n_A is the total number of observations above the detection limit ($n_A = \#(A)$).

Defining $\nu_i = (lny_i - x_i'\beta)/\sigma, w_i = (lnd_i - x_i'\beta)/\sigma$

and $Q(z) = \phi(z)/\Phi(z)$ where $\phi(z) = \frac{1}{\sqrt{2\pi}}e^{-z^2/2}$, the score components can be written thus

$$U_\sigma = \frac{\partial lnL}{\partial \sigma} = -\frac{n_A}{\sigma} + \frac{1}{\sigma}\sum_{i\epsilon A} v_i^2 - \frac{1}{\sigma}\sum_{i\epsilon B} \omega_i Q(\omega_1) \tag{2.7}$$

$$U_{\beta_j} = \frac{\partial lnL}{\partial \beta_j} = \frac{1}{\sigma}\sum_{i\epsilon A} x_{ji} v_i - \frac{1}{\sigma}\sum_{i\epsilon B} x_{ji} Q(\omega_i) \tag{2.8}$$

for $j = 0, 1, \ldots, p$.

The maximum likelihood estimates (MLEs) of σ and β are the roots of the score components. The resulting $(p+2) \times (p+2)$ information matrix I has the components

$$
\begin{aligned}
I_{\sigma\sigma} &= -\frac{\partial^2 lnL}{\partial\sigma^2} = \frac{n_A}{\sigma^2} - \frac{3}{\sigma^2}\sum_{i\epsilon A} v_i^2 \\
&\quad + \frac{1}{\sigma^2}\sum_{i\epsilon B} \omega_i Q(\omega_i)(2 - \omega_i^2 - \omega_i Q(\omega_i))
\end{aligned} \tag{2.9}
$$

$$
\begin{aligned}
I_{\sigma\beta j} &= -\frac{\partial^2 lnL}{\partial\sigma\beta_j} = -\frac{2}{\sigma^2}\sum_{i\epsilon A} x_{ji} v_i \\
&\quad + \frac{1}{\sigma^2}\sum_{i\epsilon B} x_{ji} Q(\omega_i)(1 - \omega_i^2 - \omega_i Q(\omega_i))
\end{aligned} \tag{2.10}
$$

$$
\begin{aligned}
I_{\beta_\kappa\beta_j} &= -\frac{\partial^2 lnL}{\partial\beta_k\partial\beta_j} = -\frac{1}{\sigma^2}\sum_{i\epsilon A} x_{ji} x_{\kappa i} \\
&\quad - \frac{1}{\sigma^2}\sum_{i\epsilon B} x_{ji} x_{\kappa i} Q(\omega_i)(\omega_i + Q(\omega_i))
\end{aligned} \tag{2.11}
$$

for $j, k = 0, 1, \ldots, p$.

6.2 Two-Station Analysis: Log-Likelihood and ML Equations

Consider the log-normal regression for the ratio of the pollutant concentration at the two stations over relevant explanatory variables. Denote by Y_{1i} and Y_{2i} the concentrations of the pollutant at FE and NOTL stations, respectively on the ith date and by $x_i = (1, x_{1i}, \ldots, x_{pi})'$ the corresponding vector of explanatory variables. Also, let d_{1i} and d_{2i} be the respective detection limits on the occasion i. The model is based on the assumption that

$$
\frac{Y_{2i}}{Y_{1i}} \sim \text{ Log-normal } (\mu_x = x_i'\beta, \sigma^2)
$$

Denote by y_{1i} and y_{2i} the observed measured concentrations. Because of the left-censoring occurring at each station, both left- and right-censored values are induced on the ratio of the concentrations, the censoring patterns of which are depicted in Table 2.1. The resulting joint log-likelihood can be written as

$$
\begin{aligned}
lnL &= -n_{11} ln\sigma - \frac{1}{2} \sum_{i\epsilon A_1 A_2} \left(\frac{ln(y_{2i}/y_{1i}) - x_i'\beta}{\sigma} \right)^2 \\
&+ \sum_{i\epsilon A_1 B_2} ln\Phi \left(\frac{ln(d_{2i}/y_{1i}) - x_1'\beta}{\sigma} \right) \\
&+ \sum_{i\epsilon A_2 B_1} ln \left(1 - \Phi \left(\frac{ln(y_{2i}/d_{1i}) - x_i'\beta}{\sigma} \right) \right)
\end{aligned}
$$

where A_1 and A_2 are the dates resulting in above-detection-limit measurements at FE and NOTL stations, repectively, and B_1 and B_2 the corresponding dates resulting in nondetects. Here n_{11} is the number of occasions where both stations produced above-detection-limit observations. Formulas for the score and information matrix components are derived similarly. Again, the MLEs of σ and β are the roots of the score components.

6.3 Numerical Solution: Newton–Raphson Iteration

The Newton–Raphson iterative procedure was found satisfactory for the calculation of the MLEs. The basic steps are as follows. Consider the log-normal regression model. Denote by $U(\theta)$ the $(p+2) \times 1$ score vector from equations (2.7) and (2.8), and by $I(\theta)$ the $(p+2) \times (p+2)$ information matrix from equations (2.9)–(2.11). The iteration step is

$$\hat{\theta}^{(\kappa)} = \hat{\theta}^{(\kappa-1)} + [I(\hat{\theta}^{(\kappa-1)})]^{-1} U(\hat{\theta}^{(\kappa-1)})$$

for $\kappa = 1, 2, \ldots$ The least-squares estimates where each left-censored observation y_i was replaced with $d_i/2$ were used as initial values, $\hat{\theta}^{(0)}$. Six to 12 iterations were sufficient to reach convergence.

Table 2.1

Censoring Patterns for the Ratio of the Concentrations

Censoring Status at FE	Censoring Status at NOTL	Censoring Status for Ratio
$A_1 : Y_1 > d_1$	$A_2 : Y_2 > d_2$	Y_2/Y_1 observed
$A_1 : Y_1 > d_1$	$B_2 : Y_2 \leq d_2$	$Y_2/Y_1 \leq d_2/y_1$
$B_1 : Y_1 \leq d_1$	$A_2 : Y_2 > d_2$	$Y_2/Y_1 \geq y_2/d_1$
$B_1 : Y_1 \leq d_1$	$B_2 : Y_2 \leq d_2$	$0 < Y_2/Y_1 < \infty$

An E-M algorithm strategy based on the approach by Aitkin

(1981) was also implemented. The key point for each iteration is to replace the 'incomplete' (i.e. censored) data by their conditional expected values, given the current parameter estimates (E step), and then use the standard least-squares formula to update the parameter estimates (M step). The procedure, although slower than the Newton-Raphson method, is simpler and works quite well.

6.4 Inferences

The inferences we consider rely on the standard asymptotic maximum likelihood result

$$I(\hat{\theta})^{1/2}(\hat{\theta}-\theta) \to N(0, I_{p+2})$$

when $n \to \infty$, provided the expected proportion of censored observation remains approximately constant (Lawless 1982). Here I_{p+2} is the $(p+2) \times (p+2)$ identity matrix. In particular, for component θ_i,

$$(\hat{\theta}_i - \theta_i)/\sqrt{I^{ii}} \to N(0,1)$$

where I^{ii} is the ith diagonal element of $I(\hat{\theta})^{-1}$. Thus, an approximate 100(1 - α) percent confidence interval for θ_i is

$$\hat{\theta}_i \pm z_{\alpha/2}\sqrt{I^{ii}}$$

Additionally, when the purpose is to test hypotheses about θ_i, for instance,

$$H_0 : \theta_i = \theta_{i,0} \ vs. \ H_1 : \theta_i \neq \theta_{i,0}$$

the corresponding test statistic is

$$z(\theta_{i,0}) = \frac{\hat{\theta}_i - \theta_{i,0}}{\sqrt{I^{ii}}}$$

which has the standard normal distribution, approximately, when H_0 is true.

6.5 Analysis Results for Individual Stations

The analysis reported here focuses on polychlorinated biphenyl (PCB, *ng/l* or *ng/g*), one of the priority toxic chemicals of concern. The methods developed, however, also apply to any other pollutant. In addition to the pollutant concentration measurements, solids concentration ($S, mg/l$), water flow (F, cfs) and date of sampling (T, *weeks from beginning of study*) were also recorded. The portion of the data we analysed contains measurements made on 469 water samples taken weekly from each station during the 10-year period from 2 April 1986 to 27 March 1996. An update in the detection limit was made around 20 May 1989. For PCB concentrations in water, it changed from 3.3 ng/l to 0.81 ng/l, while for PCB concentrations in solids it changed from 77 ng/g to 89 ng/g.

A program was written in Splus (1999) for Unix to carry out all the computations. The program was run for the PCB concentrations in each medium (water, solids) and each station (FE, NOTL). We took the log-flow, log-solid concentration and time as covariates. Thus, the generic form of the log-normal regression model fitted is

$$lnPCB = \beta_0 + \beta_1(lnF) + \beta_2(lnS) + \beta_3 T + \epsilon$$

where $\epsilon \sim N(0, \sigma^2)$. The result of the fit for the water measurements and some inferences are depicted in Table 2.2.

Table 2.2

Results of the Log-normal Regression Fit for PCB in Water at FE.

Regressor	MLE	Standard Error	$z(0)$	P-value	95 percent C.I.
Intercept	-4.657	5.260	-0.89	0.376	(-14.966, 5.652)
Log-Flow	0.430	0.428	1.00	0.315	(-0.408, 1.268)
Log-Solids	-0.095	0.035	-2.67	0.008	(-0.164, -0.025)
Time	-0.0019	0.0003	-7.07	0.000	(-0.0025, -0.0014)

$\hat{\sigma} = 0.4946$; C.I. = Confidence Interval

Both the log-solid concentration and time appear to have significant negative associations with the log-PCB concentration. The latter association, which is the stronger one, has implications for the objectives discussed in Section 3. It is indicative of a decrease in the concentration of PCB in the water entering the Niagara River over time. For an average log-flow and log-solid concentration, the estimated mean PCB concentration from the fitted regression model is 1.647 ng/l at the beginning of the sampling period (2 April 1986) and 0.677 ng/l at the end of the sampling period (27 March 1996). Thus, an estimated decrease of 41.1 percent in PCB in water occurred. This is a little lower than the 50 percent targeted reduction over that time (see Objective 3 in Section 3).

Table 2.3 contains a summary of the log-normal regression fits for PCB performed on single stations. The negative associations noted in water at FE between log-PCB and log-solids and time are shown consistently in water at NOTL and in solids at both stations with the association with time appearing the strongest.

6.6 Analysis Results for Concentration Ratio at the two Stations

For the log-normal regression of the ratio of the pollutant concentrations at the two stations for each medium (water, solids), the logarithm of the flow ratio, the logarithm of the solid concentration ratio and time were considered as covariates. Thus, the generic form of the log-normal regression model fitted is

$$ln\frac{PCB_2}{PCB_1} = \beta_0 + \beta_1 ln\frac{F_2}{F_1} + \beta_2 ln\frac{S_2}{S_1} + \beta_3 T + \epsilon$$

where $\epsilon \sim N(0, \sigma^2)$. The results of the fit for the log-ratio of PCB in water are presented in Table 2.4.

Note that only time appears to have a significant association with PCB and that this association is positive. This finding, again, is relevant for the objectives discussed in Section 3. It is indicative of an increase over time in the amount of PCB dumped directly into the river from sites in the vicinity.

The results for the log-ratio of PCB in solids are presented in Table 2.5. Here, both the log-ratio of solid concentration and time appear to have significant negative associations with the log-ratio of PCB concentration. Again, the latter stronger association, signals a decrease in PCB in the solids entering the river directly over time.

Table 2.3

Summary of MLEs for the Log-normal Regression Fits for PCB in Water and Solids at Each Station

	Intercept	Log-Flow	Log-Solids	Time	$\hat{\sigma}$
Water-FE	-4.657 (-0.89)	0.430 (1.00)	-0.095* (-2.67)	-0.0019* (-7.07)	0.4946
Water-NOTL	-16.472* (-2.75)	1.372* (2.82)	-0.103* (-3.13)	-0.001* (-3.27)	0.5454
Solids-FE	-7.688 (-1.28)	1.14* (2.34)	-0.574* (-10.77)	-0.0039* (-10.73)	0.6955
Solids-NOTL	6.042 (0.67)	-0.075 (-0.10)	-0.364* (-5.24)	-0.0045* (-6.51)	0.7952

($z(0) = \hat{\beta}_i/\sqrt{I^{ii}}$ in brackets)
*Indicates a significant coefficient

Table 2.4

Results of the Log-normal Regression Fit for the Ratio of PCB Concentrations in Water

Regressor	MLE	Stand. Error	$z(0)$	P-value	95 percent C.I.
Intercept	-0.320	0.173	-1.84	0.065	(-0.659, 0.020)
$Log(F_2/F_1)$	-0.275	1.487	-0.18	0.853	(-3.188, 2.639)
$Log(S_2/S_1)$	-0.050	0.080	-0.63	0.529	(-0.206, 0.106)
Time	0.0014	0.0006	2.38	0.017	(0.0002, 0.0025)

$\hat{\sigma} = 0.7991$; C.I. = Confidence Interval

A sequence plot of the PCB data suggested the apparent presence of a few outlying PCB concentrations. The log-transformation mitigated the visual effect of these outliers. To examine their effect on the log-normal regression analysis, the model was refitted excluding the outlying observations. The overall conclusions were unaffected. Only minor differences in the regression coeffcient estimates were noted with the new analysis yielding slightly smaller estimated standard errors for the regression coefficient estimates.

Table 2.5
Results of the Log-normal Regression Fit for the Ratio of PCB Concentrations in Solids

Regressor	MLE	Standard Error	$z(0)$	P-value	95 percent C. I.
Intercept	-1.035	0.198	-5.23	0	(-1.423, -0.647)
$Log(F_2/F_1)$	-2.080	2.669	-0.78	0.436	(-7.311, 3.150)
$Log(S_2/S_1)$	-0.427	0.156	-2.75	0.006	(-0.732, -0.122)
Time	-0.0031	0.0010	-3.09	0.002	(-0.0050, -0.0011)

$\widehat{\sigma} = 1.2452$; C.I. = Confidence Interval

7. Regression for Transformed Censored Data

7.1 *Maximum Likelihood Estimation (MLE) and Inference*

According to the notation used in Section 6, consider the regression model under transformation from Section 5 for data from a single station. Using equations (2.4) and (2.5), the joint log-likelihood of $\theta = (\lambda, \sigma, \beta_0, \beta_1, \dots, \beta_p)$ can be written as

$$\begin{aligned} lnL &= -n_A ln\sigma + (\lambda - 1)\sum_{i\epsilon A} lny_i \\ &\quad -\frac{1}{2}\sum_{i\epsilon A}\left(\frac{(y_i^\lambda - 1)/\lambda - x_i'\beta}{\sigma}\right)^2 \\ &\quad +\sum_{i\epsilon B} ln\Phi\left(\frac{d_i^\lambda - 1)/\lambda - x_i'\beta}{\sigma}\right) \end{aligned}$$

Defining $t_i = (y_i^\lambda - 1)/\lambda, u_i = (d_i^\lambda - 1)/\lambda$, $v_i = (t_i - x_i'\beta)/\sigma$ and $\omega_i = (u_i - x_i'\beta)/\sigma$, the score components can be written as

$$U_\lambda = \frac{\partial lnL}{\partial \lambda} = \sum_{i\epsilon A} lny_i - \frac{1}{\sigma}\sum_{i\epsilon A} v_i T_i' + \frac{1}{\sigma}\sum_{i\epsilon B} Q(\omega_i) U_i'$$

$$U_\sigma = \frac{\partial lnL}{\partial \sigma} = -\frac{n_A}{\sigma} + \frac{1}{\sigma}\sum_{i\epsilon A} v_i^2 - \frac{1}{\sigma}\sum_{i\epsilon B} \omega_i Q(\omega_i)$$

$$U_{\beta_j} = \frac{\partial lnL}{\partial \beta_j} = \frac{1}{\sigma}\sum_{i\epsilon A} x_{ji} v_i - \frac{1}{\sigma}\sum_{i\epsilon B} x_{ji} Q(\omega_i)$$

for $j = 0, 1, \ldots, p$, where

$$t_i' = \frac{(\lambda ln y_i - 1)y_i^\lambda + 1}{\lambda^2}$$

and

$$u_i' = \frac{(\lambda ln d_i - 1)d_i^\lambda + 1}{\lambda^2}$$

Expressions for the components of the information matrix I can be derived similarly.

Paralleling the log-normal regression analysis of Section 6, consider the normal regression on the Box-Cox transformation of the pollutant concentration ratio for the two stations. The censoring patterns for this ratio are as in Table 2.1. Note that both left- and right-censored observations are involved. According to the notation used in Section 6 and assuming that

$$\frac{(Y_{2i}/Y_{1i})^\lambda - 1}{\lambda} \sim N(\mu_x = x'\beta, \sigma^2),$$

the joint log-likelihood of θ is

$$\begin{aligned} lnL = & -n_{11} ln\sigma + (\lambda - 1)\sum_{i\epsilon A_1 A_2} ln\frac{y_{2i}}{y_{1i}} \\ & -\frac{1}{2}\sum_{i\epsilon A_1 A_2}\left(\frac{((y_{2i}/y_{1i})^\lambda - 1)/\lambda - x_i'\beta}{\sigma}\right)^2 \\ & + \sum_{i\epsilon A_1 B_2} ln\Phi\left(\frac{((d_{2i}/y_{1i})^\lambda - 1)/\lambda - x_i'\beta}{\sigma}\right) \\ & + \sum_{i\epsilon A_2 B_1} ln\left(1 - \Phi\left(\frac{((y_{2i}/d_{1i})^\lambda - 1)/\lambda - x_i'\beta}{\sigma}\right)\right) \end{aligned}$$

Formulas for the score and information matrix components are derived readily.

The calculation of the MLEs can be carried out simultaneously by Newton-Raphson as described in Section 6. Alternatively, one could compute the MLE of (σ, β) for a given value of λ using the program discussed in Section 6 with $(y_i^\lambda - 1)/\lambda$ in place of lny_i and $(d_i^\lambda - 1)/\lambda$ in place of lnd_i and then profiling λ as described in the next section.

7.2 Inferences and the Profile Likelihood of λ

Approximate inferences for each of the components of $\theta = (\lambda, \sigma, \beta_0, \beta_1, \ldots, \beta_p)'$ can be obtained using the asymptotic procedure outlined in Section 6. They require the MLEs of the parameters and the observed information matrix. Approximate inferences for the transformation parameter λ can also be derived via the likelihood ratio statistic. These involve the profile likelihood of λ.

For a given value of λ, denote by $\tilde{\sigma}(\lambda)$ and $\tilde{\beta}(\lambda)$ the restricted MLEs of σ and β, respectively, That is, $\tilde{\sigma}(\lambda)$ and $\tilde{\beta}(\lambda)$ are the values of σ and β that maximize $L(\lambda, \sigma, \beta)$ for λ fixed. Note that $\tilde{\sigma}(\lambda)$ and $\tilde{\beta}(\lambda)$ can be calculated using the program from Section 6. The profile likelihood (PL) of λ is

$$PL(\lambda) = L(\lambda, \tilde{\sigma}(\lambda), \tilde{\beta}(\lambda)), -\infty < \lambda < \infty$$

Clearly, maximizing $PL(\lambda)$ yields the maximum likelihood estimator of λ, $\hat{\lambda}$, and the maximum likelihood estimators of σ and β, namely $\hat{\sigma} = \tilde{\sigma}(\hat{\lambda})$ and $\hat{\beta} = \tilde{\beta}(\hat{\lambda})$. The relative profile likelihood (RPL) of λ is

$$RPL(\lambda) = \frac{PL(\lambda)}{PL(\hat{\lambda}}, -\infty < \lambda < \infty$$

The profile likelihood itself could be used for inference, for example in the form of intervals of highest profile likelihood for λ. For instance, the 100γ percent profile likelihood for λ is defined by

$$PLI_\gamma = \{\lambda : RPL(\lambda) \geq \gamma\}.$$

Every point λ inside PLI_γ has the property such that there are values of σ and β for which the joint likelihood of (λ, σ, β) in the light of the observed data is as large or larger than $\gamma L(\hat{\lambda}, \hat{\sigma}, \hat{\beta})$.

Alternatively, $RPL(\lambda)$ can in fact be converted to a likelihood ratio statistic for λ. From asymptotic theory we know that

$$-2lnRPL(\lambda) \approx \chi^2_{(1)}$$

for large samples, provided the expected proportion of censored observations remains approximately constant. Thus, an approximate $100(1-\alpha)$ percent confidence interval for λ is

$$CI_{i-\alpha} = \{\lambda : -2lNRPL(\lambda) \leq \chi^2_{\alpha,1}\}$$

Note that $CI_{1-\alpha} = PLI_{\gamma}$ where $\dot{\gamma} = exp(-\chi^2_{\alpha,1}/2)$. Thus, the 90 percent, 95 percent and 99 percent approximate confidence intervals are 25.85 percent, 14.65 percent and 3.62 percent profile likelihood intervals, respectively.

Of particular interest is the testing of

$$H_0 : \lambda = 0 \;\; vs. \;\; H_1 : \lambda \neq 0.$$

Since $\lambda = 0$ corresponds to the log-normal regression model, this test is in fact a goodness-of-fit test of the log-normal regression model within the Box-Cox family of transformation models. This can be done using the z approach or the likelihood ratio statistic approach.

7.3 Analysis Results for Individual Stations

We considered the regression of the transformed PCB concentration for each medium (water, solids) and for each station (FE, NOTL), taking log-flow, log-solid concentration and time as covariates. The generic form of the regression model under transformation that fitted is

$$\frac{PCB^{\lambda} - 1}{\lambda} = \beta_0 + \beta_1(lnF) + \beta_2(InS) + \beta_3 T + \epsilon$$

where $\epsilon \sim N(0, \sigma^2)$. The results for PCB in water at NOTL are presented in Table 2.6.

All the covariates have a significant linear effect on transformed PCB at this station, the effects being positive for log-flow and negative for log-solid concentration and time. Comparing these results with the data for log-flows in Table 2.2, we see an overall agreement with the log-normal regression analysis, with the strength of the significance being slightly higher for the regression-under-transformation results due to smaller estimated standard errors. How-

Table 2.6
Results of Regression-under-transformation Fit for PCB in Water at NOTL.

Regressor	MLE	Standard Error	$z(0)$	P-value	95 percent C. I.
Intercept	-13.558	4.492	-3.02	0.003	(-22.362, -4.755)
Log-Flow	1.127	0.365	3.09	0.002	(0.411, 1.843)
Log-Solids	-0.080	0.025	-3.23	0.001	(-0.128, -0.031)
Time	-0.0008	0.0002	-3.28	0.001	(-0.0012, -0.0003)

$\hat{\sigma} = 0.4110, \hat{\lambda} = -0.5633$, 95 percent C.I. for $\lambda : (-0.7944, -0.3323)$

ever, note that the value $\lambda = 0$ lies well outside the approximate 95 percent confidence interval (C.I.) for λ; this indicates a statistically significant deviation from the log-normal regression model at the 5 percent level of significance.

A summary of the regression-under-transformation fits for PCB is presented in Table 2.7. Time is the most significant predictor, having a negative effect on PCB, thus indicating a decrease of PCB concentrations over time. Comparing Table 2.7 with Table 2.3, we notice an overall agreement with the results from log-normal regression. Note however that the log-normal regression model ($\lambda = 0$) does not appear to provide an adequate fit for the PCB concentrations in water at NOTL and for solids at FE.

Further insight into the estimation of the transformation parameter λ is provided by Figure 2.2, where plots of the relative profile likelihood (RPL) of λ for PCB are depicted. Note that all the plots show a fair amount of symmetry. The horizontal line plotted is determined by the two values of λ for which $RPL(\lambda) = 0.1465$. These values specify the approximate 95 percent confidence interval for λ from the likelihood ratio statistic. The actual numerical values of these confidence intervals are reported in Table 2.8. Note that due to the near symmetry of the profile likelihood, these intervals are in reasonable agreement with those derived from the asymptotic normality of $\hat{\lambda}$ (Table 2.7).

Table 2.7

Summary of MLEs for the Regression-under-Transformation Fits for PCB in Water and Solids at each Station

	Intercept	Log-Flow	Log-Solids	Time	$\widehat{\sigma}$
Water-FE	-3.612 (-0.76)	0.338 (0.88)	-0.088* (-2.76)	-0.0017* (-6.17)	0.4454
	$\widehat{\lambda}$ = -0.2625 $z(\lambda = 0)$ = -1.89 95 percent $C.I.$ = (-0.5352, 0.0101)				
Water-NOTL	-13.558* (-3.02)	1.127* (3.09)	-0.080* (3.23)	-0.0008* (-3.28)	0.4110
	$\widehat{\lambda}$ = -0.5633 $z(\lambda = 0)$ = -4.7786 95 percent $C.I.$ = (-0.7944, -0.3323)				
Solids-FE	0.657 (1.22)	0.112 (1.86)	-0.049* (-2.24)	-0.0003* (-2.23)	0.0543
	$\widehat{\lambda}$ = -0.4708 $z(\lambda = 0)$ = -5.5056 95 percent $C.I.$ = (-0.6383, -0.3032)				
Solids-NOTL	2.631 (1.24)	-0.009 (-0.06)	-0.075 (-0.99)	-0.0009 (-0.99)	0.1612
	$\widehat{\lambda}$ = -0.3168 $z(\lambda = 0)$ = -1.5843 95 percent $C.I.$ = (-0.7088, 0.0751)				

($z(0) = \hat{\beta}_i/\sqrt{I^{ii}}$ in brackets)
* Indicates a significant coefficient

Note that for solids at NOTL, the profile likelihood is incomplete. This was due to a numerical difficulty we encountered with the convergence of the Newton-Raphson procedure. Several initial values were used but convergence did not occur. This case requires further examination, perhaps through a different procedure.

7.4 Numerical Results for Pollutant Concentration Ratios at the Two Stations

Regression under transformation was considered for the ratio of the pollutant concentrations at the two stations. The logarithm of the flow ratio and the logarithm of the solid concentration and time were taken as covariates. Thus, we fitted the model

$$\frac{(PCB_2/PCB_1)^\lambda - 1}{\lambda} = \beta_0 + \beta_1 ln\frac{F_2}{F_1} + \beta_2 ln\frac{S_2}{S_1} + \beta_3 T + \epsilon$$

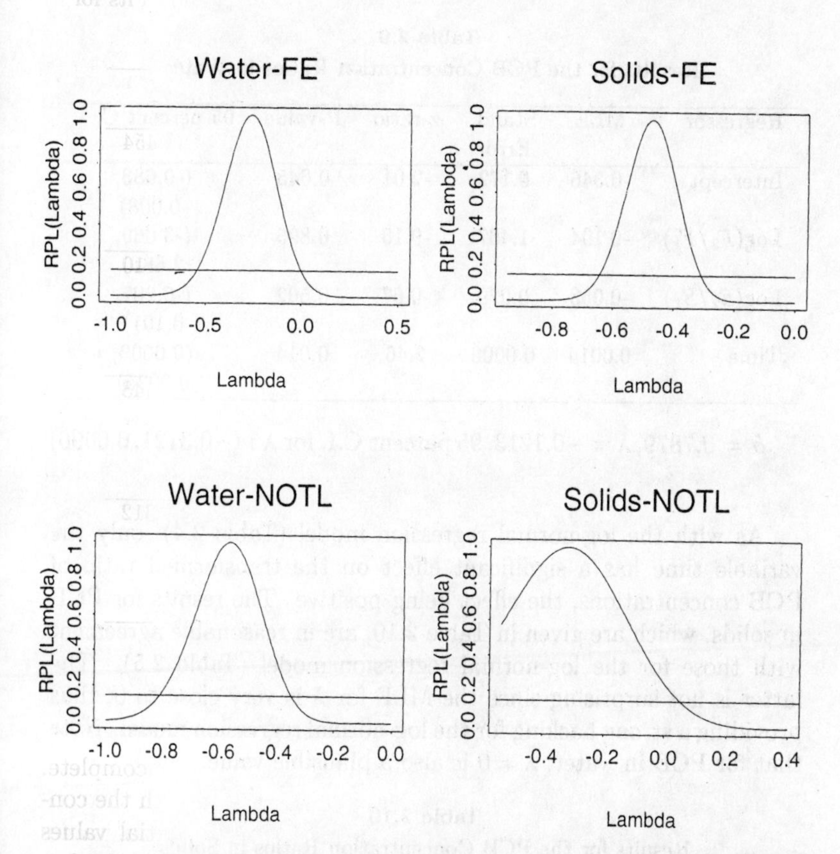

Figure 2.2: Plots of the Relative Profile Likelihood (RPL) of λ for PCB

Table 2.8

Approximate 95 percent Confidence Intervals for λ for PCB Calculated from the Likelihood Ratio Statistic

	95 percent C. I.
Water-FE	(-0.5468, 0.0002)
Water-NOTL	(-0.8085, -0.3465)
Solids-FE	(-0.6442, -0.3109)
Solids-NOTL	(N.A.,0.0521)

where $\epsilon \sim N(0, \sigma^2)$. Table 2.9 contains the results for the PCB concentration ratios in water.

Table 2.9

Results for the PCB Concentration Ratios in Water

Regressor	MLE	Stand. Error	z-ratio	P-value	95 percent C. I.
Intercept	-0.346	0.172	-2.01	0.045	(-0.683, -0.008)
$\text{Log}(F_2/F_1)$	-0.194	1.467	-0.13	0.895	(-3.069, 2.681)
$\text{Log}(S_2/S_1)$	-0.053	0.079	-0.67	0.502	(-0.207, 0.101)
Time	0.0014	0.0006	2.46	0.014	(0.0003, 0.0025)

$\hat{\sigma} = 0.7879, \hat{\lambda} = -0.1213$, 95 percent C.I. for $\lambda : (-0.3121, 0.0696)$

As with the log-normal regression model (Table 2.4), only the variable time has a significant effect on the transformed ratio of PCB concentrations, the effect being positive. The results for PCB in solids, which are given in Table 2.10, are in reasonable agreement with those for the log-normal regression model (Table 2.5). The latter is not surprising since the MLE for λ is very close to 0, thus providing a strong backing for the log-normal regression model. Note that for PCB in water, $\lambda = 0$ is also a plausible value.

Table 2.10

Results for the PCB Concentration Ratios in Solids

Regressor	MLE	Standard Error	z-ratio	P-value	95 percent C.I
Intercept	-0.976	0.198	-4.93	0.000	(-1.364, -0.588)
$\text{Log}(F_2/F_1)$	-1.864	2.530	-0.74	0.461	(-6.822, 3.095
$\text{Log}(S_2/S_1)$	-0.403	0.149	-2.70	0.007	(-0.695, -0.110)
Time	-0.0029	0.001	-2.97	0.003	(-0.0047, -0.0010)

$\hat{\sigma} = 1.1786, \hat{\lambda} = 0.0915$, 95 percent C.I. for $\lambda : (-0.1204, 0.3034)$

8. Conclusions and Future Work

Our analysis focused on the PCB concentrations reported in the Niagara River pollutant data for the 10-year period from 2 April 1986 to 27 March 1996. The two main findings in relation to the objectives of Section 3 are:

(1) A decrease, over time, in PCB concentration, both in water and in solids, in the samples taken at each of the two stations (FE, NOTL).

(2) An increase, over time, in the ratio of PCB concentrations at the two stations; this suggests that the concentration of the pollutant at FE is decreasing faster than that at NOTL.

Finding (1) suggests that increasingly cleaner water is entering the Niagara River while finding (2) suggests that the levels of PCB reaching the river over its stretch from FE to NOTL is growing over time. Note that the latter may not necessarily be due to a steady increase in the dumping of PCB during this period but may rather be the result of the slow pace with which the PCB makes its way to the river from sources whose peak pollution activity may have occurred years ago.

The above results are shown consistently by both approaches discussed in this paper, namely the log-normal regression model and regression under transformation. This suggests that although our regression–under–transformation analysis casts some doubts on the adequacy of the log-normal model in some cases, the standard practice of treating pollutant concentrations as log-normally distributed would not mislead the practitioner grossly in the analysis of PCB concentrations for the Niagara River pollutant data.

Our findings for the log-normal regression of PCB concentrations for individual stations are in agreement with those of El-Shaarawi and Al-Ibrahim (1996) who fitted this model to the same data, accounting for nondetects in the same way. Our methods, however, go beyond this analysis in two ways. One is by considering the ratio of pollutant concentrations at the two stations, thus yielding a paired analysis of the data. El-Shaarawi and Al-Ibrahim (1996) worked with the ratio of estimated average concentrations from individual station analyses instead. The other direction is by considering the regression under transformation. This approach appears to be new in the regression analysis of pollutant concentration data.

The inferential methods discussed are based on asymptotic results and thus are only approximate. It would be of practical interest to characterize their accuracy, particularly in relation to the prevalence of nondetects (censored observations). Since exact distributional results appear difficult, even for i.i.d. censored samples, simulations may be the only alternative.

Considering non-normal error distributions in the regression model for the log-transformed data may provide a useful alternative to the Box-Cox transformation approach taken here. El-Shaarawi and Viveros (1997) discussed several aspects of this approach for complete environmental data. Its performance under the presence of nondetects has not been explored yet.

Our analyses assume that the measured pollutant concentrations are independent of date. This assumption, however, is likely not valid for many pollutant concentration data taken at the same geographical location over time. The most natural way to account for possible serial correlation is by considering time-series models. While these models have been well-studied for complete data, little is known about how to fit them and how well they perform when censored observations are present. The yet unpublished work of Newcombe (1994) may be a good place to start.

REFERENCES

Aitkin, M. (1981), 'A Note on the Regression Analysis of Censored Data', *Technometrics*, 23, pp. 161-3.

Allan, R.J., A. Mudroch and A. Sudar (1983), 'An Introduction to the Niagara River/Lake Ontario Pollution Problem', *Journal of Great Lakes Research*, 9, pp. 111-7.

Box, G.E.P. and D.R. Cox (1964), 'An Analysis of Transformations' (with discussion), *Journal of the Royal Statistical Society*, B39, pp. 211-52.

COA (1981), 'Canada-Ontario Agreement on Great Lakes Water Quality', Environmental Baseline Report of the Niagara River, Ontario: Environment Canada and Ministry of the Environment.

Currie, L.A. (1988), *Detection in Analytical Chemistry: Importance, Theory and Practice*, Washington DC: American Chemical Society.

Eadie, B.J. and A. Roberton (1976), 'A Carbon Budget for Lake Ontario', *Journal of Great Lakes Research*, 2, pp. 307-23.

El-Shaarawi, A.H. (1989), 'Inferences About the Mean from Censored Water Quality Data', *Water Resources Research*, 25, pp. 685-90.

El-Shaarawi, A.H. and A.H. Al-Ibrahim (1996), 'Final Report: Trend Analysis and Maximum Likelihood Estimation of Niagara River Data (1986-1984)', Technical Report, National Water Research Institute and McMaster University.

El-Shaarawi, A.H. and R. Viveros (1997), 'Inference about the Mean in Log-Regression With Environmental Applications', *Environmetrics*, 8, pp. 569-82.

Frank, R., R.L. Thomas, M. Holdrinet, A.L.W. Kemp and H.E. Braun (1979), 'Organochlorine Insecticides and PCB in Surficial Sediments (1968) and Sediment Cores (1976) from Lake Ontario', *Journal of Great Lakes Research*, 9, pp. 183-89.

Kaminsky, R., K.L.E. Kaiser and R.A. Hites (1983), 'Fates of Organic Compounds From Niagara Falls Dumpsites in Lake Ontario', *Journal of Great Lakes Research*, 9, pp. 183-89.

Kemp, A.L.W. and N.S. Harper (1976), 'Sedimentation Rates and a Sedimentation Budget for Lake Ontario', *Journal of Great Lakes Research*, 2, pp. 324-40.

Haile, C. L. (1977), 'Chlorinated Hydrocarbons in the Lake Ontario and Lake Michigan Ecosystems', Dissertation, University of Wisconsin-Madison.

Holdrinet, M. V., R. Frank, R.L. Thomas and L.J. Hetling (1978), 'Mirex in the Sediments of Lake Ontario', *Journal of Great Lakes Research*, 4, pp. 69-74.

Lambert, D., B. Peterson, and I. Terpenning (1991), 'Nondetects, Detection Limits, and Probability of Detection', *Journal of the American Statistical Association*, 86, pp. 266-77.

Lawless, J.F. (1982), *Statistical Models and Methods for Lifetime Data*, Wiley: New York.

Newcombe, P.A. (1994), Issues in the Use of Double Exponential Autoregressive (1) Models, unpublished Ph.D. thesis, Department of Statistics and Actuarial Science, University of Waterloo.

Niagara River Analytical Protocol (1987), A Joint EC, USEPA, NYSDEC, and MOE Report, Ontario: Environmental Canada.

NRTC (1984), Report of The Niagara River Toxics Committee, *Joint Report by the Canadian and United States Governments*, Ontario: Environmental Canada.

Ott, W.R. (1995), *Environmental Statistics and Data Analysis*, Boca Raton: CRC Press.

Ott, W.R. (1990), 'A Physical Explanation of the Log-normality of Pollutant Concentrations', *Journal of the Air and Waste Management Association*, 40, pp. 1378-83.

Schmoyer, R.L., J.J. Beauchamp, C.C. Brandt and F.O. Hoffman Jr. (1996), 'Difficulties With the Lognormal Model in Mean Estimation and Testing', *Environmental and Ecological Statistics*, 3, pp. 81-97.

Shumway, R.H., A.S. Azri and P. Johnson (1989), 'Estimating Mean Concentrations Under Transformation for Environmental Data With Detection Limits', *Technometrics*, 31, pp. 347-57.

Splus (1999), SPLUS version 5.0 Release 2 for SunOS 5.5, Seattle, USA: Insightful Corporation.

Vincent, J. and A. Franzen (1982), *An Overview of Environmental Pollution in the Niagara Frontier, New York.* USEPA, National Enforcement Investigation Center, Denver, Colorado.

3

A Space-time State-space Modelling of Tokyo Bay Pollution

Nobuhisa Kashiwagi • Katsuyuki Ninomiya • Haruo Ando • Hisako Ogura

A state-space model for space-time seasonal adjustment is proposed to analyse Tokyo Bay pollution data. The model is constructed in steps. First, a model for space-time smoothing is derived by combining a time-series model and a spatial model using a generalized inverse technique. Then, the model is modified to treat missing values. These models are applied to artificial data to check their performance. Finally, a model for space-time seasonal adjustment is derived and is applied to Tokyo Bay pollution data.

1. Introduction

In Tokyo Bay, water quality is measured by the relevant local governments every month at scores of monitoring points for administrative purposes. We are carrying out a plan of collecting and screening the measured values reported for the past two decades in order to use them for scientific purposes. This plan has been completed for the data obtained during April 1985 to March 1990. It has been confirmed from the collected data that most of the measured quantities change seasonally at most of the monitoring points (Kashiwagi 1997 and Ninomiya *et al.* 1997, 1996a, 1996b). Therefore, the data we

---

Supported in part by Grant-in-Aid for Scientific Research (C) 11680-332 from the Japanese Ministry of Education, Science, Sports and Culture.

attempt to analyze are space-time data which involve seasonal variation. In this paper, we propose a state-space model for space-time seasonal adjustment to analyze the Tokyo Bay data. The natural environment is more or less affected by the revolution of the earth. Therefore, many applications of space-time seasonal adjustment may be found in a study on the changes in the natural environment in a wide area over many years.

The space-time seasonal adjustment is based on a time-series seasonal adjustment method. Time-series seasonal adjustment has been discussed mainly in econometrics, and empirical methods based on moving average such as X-11 (Shiskin *et al.* 1967) have been widely used. However, as the basis of space-time seasonal adjustment, model-based methods are preferable because they enable us to deal with space-time variation on purpose.

To eliminate ambiguities involved in X-11, many authors have discussed various model-based methods (see, for example, Box and Jenkins 1970, Hillmer and Tiao 1982, Harvey 1989, and Ozaki and Thomson 1994). Among them, in the preliminary analysis of the Tokyo Bay data, we have employed the Bayesian seasonal adjustment method proposed by Akaike (1980), who assumed the following model to decompose a time-series of monthly observations $\mathbf{y} = (y_1, \ldots, y_n)'$ to trend $\mathbf{t} = (t_1, \ldots, t_n)'$, seasonal $\mathbf{s} = (s_1, \ldots, s_n)'$ and residual $\mathbf{u} = (u_1, \ldots, u_n)'$ components.

$$p(\mathbf{y} \mid \mathbf{t}, \mathbf{s}, \sigma^2): \quad y_k = t_k + s_k + u_k, \quad u_k \sim NID(0, \sigma^2),$$

$$k = 1, \ldots, n$$

$$p(\mathbf{t} \mid \alpha^2): \quad t_k - 2t_{k-1} + t_{k-2} \sim NID(0, \alpha^2),$$

$$k = 3, \ldots, n$$

$$p(\mathbf{s} \mid \beta^2, \gamma^2): \quad s_k - s_{k-12} \sim NID(0, \beta^2),$$

$$k = 13, \ldots, n$$

and

$$\sum_{l=0}^{11} s_{k-l} \sim NID(0, \gamma^2), \quad k = 12, \ldots, n$$

where $p(\cdot)$'s denote the corresponding densities. Then, he proposed to use the posterior density

$$p(\mathbf{t}, \mathbf{s} \mid \mathbf{y}, \sigma^2, \alpha^2, \beta^2, \gamma^2) \propto$$
$$p(\mathbf{y} \mid \mathbf{t}, \mathbf{s}, \sigma^2) p(\mathbf{t} \mid \alpha^2) p(\mathbf{s} \mid \beta^2, \gamma^2) \tag{3.1}$$

to estimate **t** and **s**, and the Bayesian likelihood

$$L(\sigma^2, \alpha^2, \beta^2, \gamma^2) = \int_{R(\mathbf{t},\mathbf{s})} p(\mathbf{y} \mid \mathbf{t}, \mathbf{s}, \sigma^2) p(\mathbf{t} \mid \alpha^2) \times p(\mathbf{s} \mid \beta^2, \gamma^2) d\mathbf{t} d\mathbf{s} \tag{3.2}$$

to estimate $\sigma^2, \alpha^2, \beta^2$ and γ^2, where $R(\mathbf{t}, \mathbf{s})$ is the support of **t** and **s**. By using this method, we have succeeded in revealing basic characteristics of water quality in Tokyo Bay (Ninomiya *et al.* 1997, 1996a, 1996b). It is not realistic to construct a space-time seasonal adjustment method by extending Akaike's method directly, because it requires an unreasonable cost to evaluate equations (3.1) and (3.2) in a large-scale problem such as space-time seasonal adjustment. In space-time seasonal adjustment, the sizes of **t** and **s** depend on both the number of observation points and the number of time periods.

A realistic way of constructing a space-time seasonal adjustment method is to use the state-space approach. Motivated by Akaike (1980), Kitagawa and Gersh (1984) discussed the state-space approach to time-series seasonal adjustment and gave a state-space representation of Akaike's model as follows:

$$p(y_k \mid \mathbf{x}_k): \quad y_k = F\mathbf{x}_k + u_k, \quad u_k \sim NID(0, \sigma^2),$$

$$p(\mathbf{x}_k \mid \mathbf{x}_{k-1}): \quad \mathbf{x}_k = G\mathbf{x}_{k-1} + \mathbf{v}_k, \quad \mathbf{v}_k \sim NID(0, H),$$

where $\mathbf{x}_k = (t_k, t_{k-1}, s_k, \ldots, s_{k-10})'$,

$$F = (1, 0, 1, 0, \ldots, 0),$$

$$G = \begin{bmatrix} 2 & -1 & & & & \\ 1 & 0 & & & & \\ & & -1 & \ldots & -1 & -1 \\ & & 1 & & & \\ & & & \ldots & & \\ & & & & 1 & \end{bmatrix}$$

and

$$H = \begin{bmatrix} \alpha^2 & & & & \\ & 0 & & & \\ & & \gamma^2 & & \\ & & & 0 & \\ & & & & \cdots \\ & & & & & 0 \end{bmatrix}$$

Once a state-space model is given, the recursive formulae, including the Kalman filter, are available for the estimation. The recursive formulae are written in the density form as:

(prediction):

$$p(\mathbf{x}_k \mid \mathbf{Y}_{k-1})$$

$$= \int_{R(\mathbf{x}_{k-1})} p(\mathbf{x}_k \mid \mathbf{x}_{k-1}) p(\mathbf{x}_{k-1} \mid \mathbf{Y}_{k-1}) d\mathbf{x}_{k-1},$$

(filtering):

$$p(\mathbf{x}_k \mid \mathbf{Y}_k) = \frac{p(y_k \mid \mathbf{x}_k) p(\mathbf{x}_k \mid \mathbf{Y}_{k-1})}{p(y_k \mid \mathbf{Y}_{k-1})},$$

$$p(y_k \mid \mathbf{Y}_{k-1}) = \int_{R(\mathbf{x}_k)} p(y_k \mid \mathbf{x}_k) p(\mathbf{x}_k \mid \mathbf{Y}_{k-1}) d\mathbf{x}_k,$$

(likelihood):

$$L(\theta) = \prod_{k=1}^{n} p(y_k \mid \mathbf{Y}_{k-1}),$$

(smoothing):

$$p(\mathbf{x}_k \mid \mathbf{Y}_n) =$$

$$\int_{R(\mathbf{x}_{k+1})} \frac{p(\mathbf{x}_{k+1} \mid \mathbf{x}_k) p(\mathbf{x}_k \mid \mathbf{Y}_k)}{p(\mathbf{x}_{k+1} \mid \mathbf{Y}_k)} p(\mathbf{x}_{k+1} \mid \mathbf{Y}_n) d\mathbf{x}_{k+1},$$

where $\mathbf{Y}_k = (y_1, \ldots, y_k)$ and in Kitagawa-Gersh's model, $\theta = (\sigma^2, \alpha^2, \gamma^2)$. The size of $\mathbf{x}_k$ does not depend on the number of time periods. Therefore, only a little computation is required. By this reason, the state-space approach has offered an attractive way to solve complex or large-scale problems (for example, Kashiwagi 1996, 1993, Kashiwagi and Yanagimoto 1992 and Kitagawa 1998, 1987).

On the other hand, to construct a space-time model based on a time-series model, it is necessary to incorporate spatial variation into the time-series model. If spatial variation is stationary, ordinary ways may be available to incorporate spatial variation (see, for example, Cressie 1991). However, Tokyo Bay has a closed shape topographically, the distance between the ocean and the innermost point of the bay is relatively long, and the big rivers drain into the bay at the inner side of the bay. Owing to these facts, water quality in Tokyo Bay varies with location having spatial trend (Ninomiya *et al.* 1997, 1996a, 1996b). To represent such non-stationary spatial variation, the discrete thin plate smoothing model (DTPSM) (Kashiwagi 1993) is useful.

The purpose of this paper is to propose a state-space model for space-time seasonal adjustment based on Kitagawa-Gersh's model and the DTPSM. However, its modelling is not straightforward. Therefore, we first investigate space-time smoothing in Section 2, and then investigate space-time seasonal adjustment in Section 3. In Section 2.1, a state-space model for space-time smoothing is derived by combining a time-series smoothing model and the DTPSM using a generalized inverse technique. In Section 2.2, the treatment of missing values is discussed. In Section 2.3, the model is modified to treat missing values in space. In Section 2.4, the results of a simulation study are shown. In Section 3.1, a state-space model for space-time seasonal adjustment is constructed based on the discussions in Section 2. In Section 3.2, a few remarks on the proposed model are given. In Section 3.3, the proposed model is applied to the Tokyo Bay data.

2. Space-time Smoothing

2.1 Basic model

Let $y_{ij}(k)$ be an observation at the (i,j)th site on a two-dimensional rectangular lattice at a time period k, where the variables $i = 1,\ldots,n_i$, $j = 1,\ldots,n_j$ and $k = 1,\ldots,n_k$. To smooth $y_{ij}(k)$s, we assume the following model:

$$y_{ij}(k) = t_{ij}(k) + u_{ij}(k), \quad u_{ij}(k) \sim NID(0,\sigma^2) \tag{3.3}$$

$$t_{ij}(k) - 2t_{ij}(k-1) + t_{ij}(k-2) \sim NID(0,\alpha^2) \tag{3.4}$$

$$4t_{ij}(k) - t_{i+1,j}(k) - t_{i-1,j}(k) - t_{i,j+1}(k)$$

$$-t_{i,j-1}(k) \sim NID(0, \beta^2) \tag{3.5}$$

where $t_{ij}(k)$ is a trend component, and it is assumed that $t_{0j}(k) = t_{1j}(k)$, $t_{n_i+1,j}(k) = t_{n_ij}(k)$, $t_{i0}(k) = t_{i1}(k)$ and $t_{i,n_j+1}(k) = t_{in_j}(k)$. The pair of equations (3.3) and (3.4) gives a time-series smoothing model for fixed (i, j), and that of equations (3.3) and (3.5) gives the DTPSM for fixed k. The assumed model can be written in the matrix form as:

$$\mathbf{y}(k) = \mathbf{t}(k) + \mathbf{u}(k), \quad \mathbf{u}(k) \sim NID(\mathbf{0}, \sigma^2 I) \tag{3.6}$$

$$\begin{bmatrix} I \\ D \end{bmatrix} \mathbf{t}(k) = \begin{bmatrix} 2I & -I \\ \mathbf{0} & \mathbf{0} \end{bmatrix} \begin{pmatrix} \mathbf{t}(k-1) \\ \mathbf{t}(k-2) \end{pmatrix} + \begin{pmatrix} \mathbf{v}_\alpha(k) \\ \mathbf{v}_\beta(k) \end{pmatrix} \tag{3.7}$$

$$\mathbf{v}_\alpha(k) \sim NID(\mathbf{0}, \alpha^2 I), \quad \mathbf{v}_\beta(k) \sim NID(\mathbf{0}, \beta^2 I)$$

where, $\mathbf{y}(k)$ and $\mathbf{t}(k)$ are column vectors of $y_{ij}(k)$s and $t_{ij}(k)$s respectively, and D is a matrix to represent the linear combinations appearing in equation (3.5). Equation (3.6) can be regarded as an observation model in the state-space approach. However, equation (3.7) cannot be a system model in its form, because the left side involves the unwelcome coefficient. We eliminate this coefficient by using the Moore-Penrose inverse. Then, we have

$$\mathbf{t}(k) = \left[2\alpha^{-2}T, \ -\alpha^{-2}T\right] \begin{pmatrix} \mathbf{t}(k-1) \\ \mathbf{t}(k-2) \end{pmatrix} + \mathbf{v}_t(k) \tag{3.8}$$

$$\mathbf{v}_t(k) \sim NID(\mathbf{0}, T), \quad T = (\alpha^{-2}I + \beta^{-2}D'D)^{-1}$$

Using equations (3.6) and (3.8), we give the basic model for space-time smoothing as:

$$\mathbf{y}(k) = F\mathbf{x}(k) + \mathbf{u}(k), \quad \mathbf{u}(k) \sim NID(\mathbf{0}, \sigma^2 I) \tag{3.9}$$

$$\mathbf{x}(k) = G\mathbf{x}(k-1) + v(k), \quad \mathbf{v}(k) \sim NID(\mathbf{0}, H) \tag{3.10}$$

where $\mathbf{x}(k) = (\mathbf{t}'(k), \mathbf{t}'(k-1))'$, $F = [I, \mathbf{0}]$,

$$G = \begin{bmatrix} 2\alpha^{-2}T & -\alpha^{-2}T \\ I & \mathbf{0} \end{bmatrix} \text{ and } H = \begin{bmatrix} T & \mathbf{0} \\ \mathbf{0} & \mathbf{0} \end{bmatrix}.$$

This is a Gaussian state-space model; the Kalman filter and the related recursive formulae are available for the estimation.

Note that equation (3.8) is not exactly equivalent to equation (3.7). However, the Moore-Penrose inverse has a close relation with the least-squares method, and therefore, it is strongly expected that equation (3.8) is not so far from equation (3.7) in practice.

2.2 Missing values in time

In the state-space approach, missing values are treated usually by skipping the corresponding filtering steps. To treat missing values involved in the Tokyo Bay data, we use this technique by rewriting equation (3.7) as:

$$J(k)\mathbf{y}(k) = J(k)\mathbf{t}(k) + J(k)\mathbf{u}(k) \tag{3.11}$$

where $J(k)$ is a matrix generated from the identity matrix by deleting the rows corresponding to missing values during a time period k. However, in a certain case, the use of equation (3.11) causes a numerical problem.

The Kalman filter for our model is given as follows:
(prediction):

$$\begin{aligned} \mathbf{x}_{k/k-1} &= G\mathbf{x}_{k-1/k-1}, \\ \Sigma_{k/k-1} &= G\Sigma_{k-1/k-1}\,G' + H \end{aligned}$$

(filtering):

$$\begin{aligned} \mathbf{x}_{k/k} &= \mathbf{x}_{k/k-1} + \Sigma_{k/k-1}F' \times \\ &\quad \left(F\Sigma_{k/k-1}F' + \sigma^2 I\right)^{-1} \times \\ &\quad \left(\mathbf{y}(k) - F\mathbf{x}_{k/k-1}\right), \\ \Sigma_{k/k} &= \Sigma_{k/k-1} - \Sigma_{k/k-1}F' \times \\ &\quad \left(F\Sigma_{k/k-1}F' + \sigma^2 I\right)^{-1}\, F\Sigma_{k/k-1} \end{aligned}$$

where $\mathbf{x}_{i/j}$ and $\sum_{i/j}$ are the conditional mean and covariance of $\mathbf{x}(i)$ given $\mathbf{y}(1), \ldots, \mathbf{y}(j)$ respectively. Roughly speaking, the value of each component of $\sum_{i/j}$ expands and shrinks through the prediction and filtering steps respectively. However, if equation (3.11) is used, the values of the covariance components corresponding to missing values do not shrink sufficiently in the filtering step. Therefore, if observations are not available consecutively at certain sites, the values of the corresponding covariance components expand infinitely, while those of the other covariance components are kept at constant levels. Then, it happens that $(F\sum_{k/k-1} F' + \sigma^2 I)^{-1}$ cannot be calculated numerically because of round-off errors, and the Kalman filter is terminated necessarily. The discussion in the next section is related to this numerical problem.

2.3 Missing values in space

The model proposed in Section 2.1 is based on a lattice, while the monitoring points in Tokyo Bay locate irregularly. If a fine lattice is used to represent Tokyo Bay, it may be allowed to assign the monitoring points to the nearest sites. However, at the same time, it happens that there are many sites which do not correspond to the monitoring points. Then, the numerical problem mentioned in the previous section arises, because no observation is available at those sites. In this section, we give a model for the case where sites without observation exist.

Let $\mathbf{t}_e(k)$ and $\mathbf{t}_m(k)$ be column vectors of $t_{ij}(k)$s at sites with observations and at sites without observation respectively . The observation model is given naturally as:

$$\mathbf{y}_e(k) = \mathbf{t}_e(k) + \mathbf{u}_e(k), \quad \mathbf{u}_e(k) \sim NID(\mathbf{0}, \sigma^2 I) \tag{3.12}$$

where $\mathbf{y}_e(k)$ is a column vector of $y_{ij}(k)$s at sites with observations. On the other hand, we assume the following model corresponding to equation (3.7).

$$\begin{bmatrix} I & \mathbf{0} \\ D_e & D_m \end{bmatrix} \begin{pmatrix} \mathbf{t}_e(k) \\ \mathbf{t}_m(k) \end{pmatrix} = \begin{bmatrix} 2I & -I \\ \mathbf{0} & \mathbf{0} \end{bmatrix} \begin{pmatrix} \mathbf{t}_e(k-1) \\ \mathbf{t}_e(k-2) \end{pmatrix} + \begin{pmatrix} \mathbf{v}_{e\alpha}(k) \\ \mathbf{v}_\beta(k) \end{pmatrix} \tag{3.13}$$

$$\mathbf{v}_{e\alpha}(k) \sim NID(\mathbf{0}, \alpha^2 I), \quad \mathbf{v}_\beta(k) \sim NID(\mathbf{0}, \beta^2 I)$$

where D_e and D_m are submatrices generated by rearranging the columns of D corresponding to $\mathbf{t}_e(k)$ and $\mathbf{t}_m(k)$ and paying the attention to the extensibility of the model; smoothness in time is not assumed for $\mathbf{t}_m(k)$. We eliminate the coefficient $\mathbf{0}$ on the left side of equation (3.13) using the Moore-Penrose inverse as in equation (3.7). Then, since the two equations derived from the first and second rows of equation (3.13) are linearly dependent, we finally obtain

$$\mathbf{t}_e(k) = \left[2\alpha^{-2}T_e, \quad -\alpha^{-2}T_e\right]\begin{pmatrix}\mathbf{t}_e(k-1) \\ \mathbf{t}_e(k-2)\end{pmatrix} + \mathbf{v}_{te}(k) \qquad (3.14)$$

$$\mathbf{v}_{te}(k) \sim NID(\mathbf{0}, T_e), \quad T_e = \{\alpha^{-2}I + \beta^{-2}D_e'(I - D_m D_m^+)D_e\}^{-1}$$

where D_m^+ is the Moore-Penrose inverse of D_m. Using equation (3.12) and equation (3.14), we give the model for the case where sites without observation exist as:

$$\mathbf{y}_e(k) = F_e\mathbf{x}_e(k) + \mathbf{u}_e(k), \quad \mathbf{u}_e(k) \sim NID(\mathbf{0}, \sigma^2 I) \qquad (3.15)$$

$$\mathbf{x}_e(k) = G_e\mathbf{x}_e(k-1) + \mathbf{v}_e(k), \quad \mathbf{v}_e(k) \sim NID(\mathbf{0}, H_e) \qquad (3.16)$$

where $\mathbf{x}_e(k) = (\mathbf{t}_e'(k), \mathbf{t}_e'(k-1))'$, $F_e = [I, \mathbf{0}]$,

$$G_e = \begin{bmatrix} 2\alpha^{-2}T_e & -\alpha^{-2}T_e \\ I & \mathbf{0} \end{bmatrix} \quad \text{and} \quad H_e = \begin{bmatrix} T_e & \mathbf{0} \\ \mathbf{0} & \mathbf{0} \end{bmatrix}.$$

The unknown quantities can be estimated by using the Kalman filter and the related recursive formulae. To estimate the remaining component $\mathbf{t}_m(k)$, we use the following interpolation:

$$\hat{\mathbf{t}}_m(k) = -D_m^+ D_e \hat{\mathbf{t}}_e(k),$$

where $\hat{\mathbf{t}}_e(k)$ is an estimate of $\mathbf{t}_e(k)$. This is derived from the spatial condition $D\mathbf{t}(k) = \mathbf{0}$ by using the Moore-Penrose inverse.

2.4 Simulation

In this section, we show the results of simulation conducted to check the performance of the proposed models.

Example 2.1 To check the performance of equations (3.9) and (3.10), we prepared 48 frames on the lattice of size 19 × 19. The model used to generate the true image is:

$$\begin{aligned} t_{ij}(k) \quad &= \quad 500\sin\left(\frac{k\pi}{24}\right)\phi(i, j \mid 7, 13, 9, 9, 0.5) \\ &\quad -500\cos\left(\frac{k\pi}{24}\right)\phi(i, j \mid 13, 7, 9, 9, -0.5), \end{aligned}$$

and the model used to generate the noisy data is

$y_{ij}(k) = t_{ij}(k) + e_{ij}(k)$, where $i, j = 1, \ldots, 19$, $k = 1, \ldots, 48$, $\phi(i, j \mid \mu_x, \mu_y, \sigma_x^2, \sigma_y^2, \rho)$ is the probability density function of the bi-

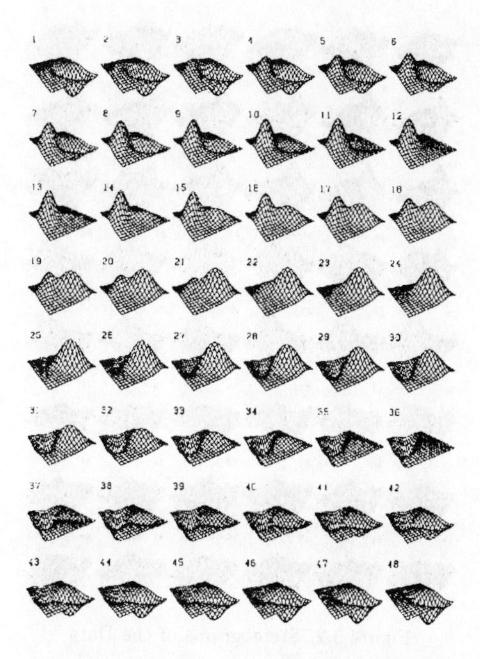

Figure 3.1: Stereograms of the True Image

variate Gaussian distribution and $e_{ij}(k)$ is a Gaussian random number with mean 0 and variance 1. Figures 3.1 and 3.2 show the stereograms of the true image and the data, respectively. To this data set, we applied equations (3.9) and (3.10). The maximum likelihood estimates obtained were $\hat{\sigma}^2 = 0.965$, $\hat{\alpha}^2 = 2.38$ and $\hat{\beta}^2 = 15.9$. The conditional means of the trend components, given all of the data obtained by using the smoothing formula, which we call the estimates of the trend components in this paper, are shown in Figure 3.3. It may be seen from the figure that the true image is successfully reconstructed from the noisy data.

Example 2.2 To simulate the case where sites without observation exist, we made a data set from the data used in the previous example by omitting the data at 80% of the sites selected randomly. Then, we applied equations (3.15) and (3.16) to this data set. The maximum likelihood estimates obtained were $\hat{\sigma}^2 = 0.950, \hat{\alpha}^2 = 12.5$ and $\hat{\beta}^2 = 15.4$. The estimates of the trend components are shown in Figure

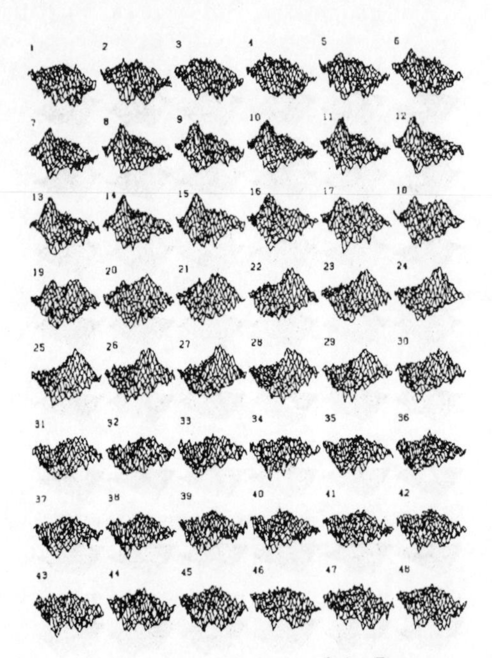

Figure 3.2: Stereograms of the Data

3.4. Here, the shape of the estimates is a little too smooth, and the two peaks are under-estimated. However, these are due to the fact that sites without observation exist. It may be said that the true image is reasonably reconstructed.

3. Space-time Seasonal Adjustment

3.1. Model for the Tokyo Bay data

In this section, we construct a state-space model for space-time seasonal adjustment based on the discussions in Section 2 considering characteristics of the Tokyo Bay data. We employ 1 km meshes to represent a Tokyo Bay area as shown in Figure 3.5, though the meshes on the land and on the tidal land are not drawn. Each mesh corresponds to a site of a lattice, and the site number starts from the northwest corner. The monitoring points, whose locations are indicated by the numbers in the figure, are assigned to the nearest

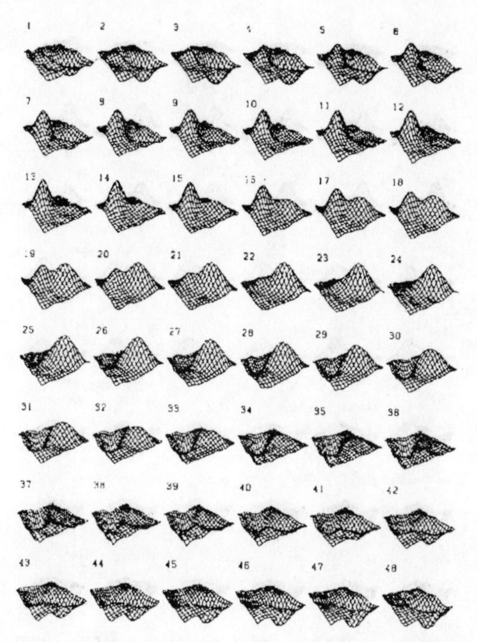

Figure 3.3: Stereograms of the Trend Estimated from all of the Data

sites. Let Ω_b be the set of the site numbers of the meshes drawn in the figure, and Ω_e be that of the monitoring points.

We assume the following model for monthly observations:

$$\{y_{ij}(k) \mid (i,j) \in \Omega_e, \quad k = 1, \ldots, n_k\}$$

$$y_{ij}(k) = t_{ij}(k) + s_{ij}(k) + u_{ij}(k)$$

$$u_{ij}(k) \sim NID(0, \sigma_{ij}^2) \tag{3.17}$$

where $t_{ij}(k)$ and $s_{ij}(k)$ are trend and seasonal components respectively. This model can be written in the matrix form as:

$$\mathbf{y}_e(k) = \mathbf{t}_e(k) + \mathbf{s}_e(k) + \mathbf{u}_e(k)$$

$$\mathbf{u}_e(k) \sim NID(\mathbf{0}, U_e) \tag{3.18}$$

where $\mathbf{y}_e(k)$, $\mathbf{t}_e(k)$ and $\mathbf{s}_e(k)$ are column vectors of $y_{ij}(k)$s, $t_{ij}(k)$s and $s_{ij}(k)$s respectively at the sites of Ω_e.

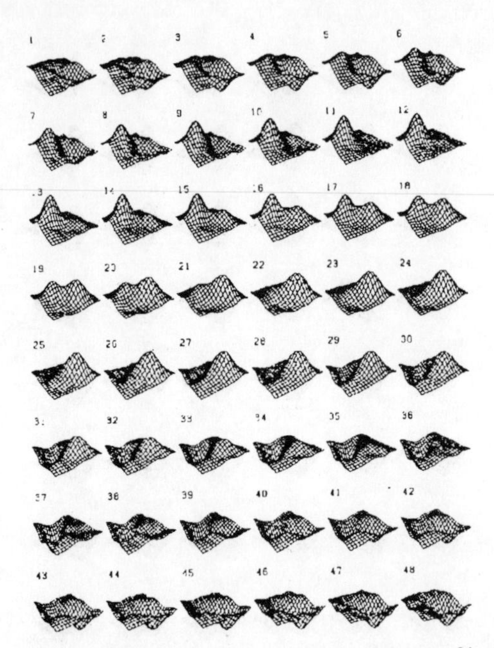

Figure 3.4: Stereograms of the Trend Estimated from 20% of the Data

On the other hand, we assume the following model for the trend components:

$$t_{ij}(k) - 2t_{ij}(k-1) + t_{ij}(k-2) \sim NID(0, \alpha_{ij}^2) \tag{3.19}$$

$$\begin{aligned} &4t_{ij}(k) - t_{i+1,j}(k) - t_{i-1,j}(k) - t_{i,j+1}(k) \\ &-t_{i,j-1}(k) \sim NID(0, \beta^2) \end{aligned} \tag{3.20}$$

where $(i,j) \in \Omega_e$ for equation (3.19) and $(i,j) \in \Omega_b$ for equation (3.20), and it is assumed that

$$t_{i\pm1,j}(k) = t_{ij}(k) \text{ for } (i \pm 1, j) \notin \Omega_b \text{ and}$$

$$t_{i,j\pm1}(k) = t_{ij}(k) \text{ for } (i, j \pm 1) \notin \Omega_b.$$

This model can be written in the matrix form as:

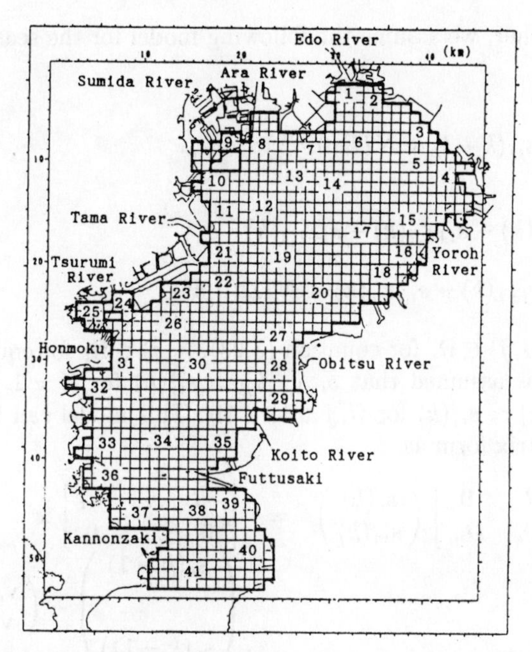

Figure 3.5: A Map of Tokyo Bay Area. The Numbers indicate the Locations of the Monitoring Points.

$$\begin{bmatrix} I & \mathbf{0} \\ D_e & D_m \end{bmatrix} \begin{pmatrix} \mathbf{t}_e(k) \\ \mathbf{t}_m(k) \end{pmatrix} = \begin{bmatrix} 2I & -I \\ \mathbf{0} & \mathbf{0} \end{bmatrix} \begin{pmatrix} \mathbf{t}_e(k-1) \\ \mathbf{t}_e(k-2) \end{pmatrix} + \begin{pmatrix} \mathbf{v}_{e\alpha}(k) \\ \mathbf{v}_\beta(k) \end{pmatrix} \tag{3.21}$$

$$\mathbf{v}_{e\alpha}(k) \sim NID(\mathbf{0}, V_{e\alpha}), \quad \text{and } \mathbf{v}_\beta(k) \sim NID(\mathbf{0}, \beta^2 I)$$

where $\mathbf{t}_m(k)$ is a column vector of $t_{ij}(k)$s at the sites of $\Omega_b - \Omega_e$, and D_e and D_m are submatrices to represent the linear combinations appearing in equation (3.20). By eliminating the coefficient on the left side of equation (3.21) as in Section 2.3, we obtain

$$\mathbf{t}_e(k) = \left[2T_e V_{e\alpha}^{-1}, \quad -T_e V_{e\alpha}^{-1}\right] \begin{pmatrix} \mathbf{t}_e(k-1) \\ \mathbf{t}_e(k-2) \end{pmatrix} + \mathbf{v}_{te}(k)$$

$$\mathbf{v}_{te}(k) \sim NID(\mathbf{0}, T_e),$$

$$T_e = \{V_{e\alpha}^{-1} + \beta^{-2} D_e'(I - D_m D_m^+) D_e\}^{-1} \tag{3.22}$$

Further, we assume the following model for the seasonal components.

$$\sum_{l=0}^{11} s_{ij}(k-l) \sim NID(0, \gamma_{ij}^2) \tag{3.23}$$

$$4s_{ij}(k) - s_{i+1,j}(k) - s_{i-1,j}(k)$$

$$-s_{i,j+1}(k) - s_{i,j-1}(k) \sim NID(0, \delta^2) \tag{3.24}$$

where $(i,j) \in \Omega_e$ for equation (3.23), $(i,j) \in \Omega_b$ for equation (3.24), and it is assumed that $s_{i\pm1,j}(k) = s_{ij}(k)$ for $(i \pm 1, j) \notin \Omega_b$ and $s_{i,j\pm1}(k) = s_{ij}(k)$ for $(i, j \pm 1) \notin \Omega_b$. This model can be written in the matrix form as:

$$\begin{bmatrix} I & \mathbf{0} \\ D_e & D_m \end{bmatrix} \begin{pmatrix} \mathbf{s}_e(k) \\ \mathbf{s}_m(k) \end{pmatrix} = \begin{bmatrix} -I & \ldots & -I \\ \mathbf{0} & \ldots & \mathbf{0} \end{bmatrix} \times \begin{pmatrix} \mathbf{s}_e(k-1) \\ \vdots \\ \mathbf{s}_e(k-11) \end{pmatrix} + \begin{pmatrix} \mathbf{v}_{e\gamma}(k) \\ \mathbf{v}_\delta(k) \end{pmatrix} \tag{3.25}$$

$$\mathbf{v}_{e\gamma}(k) \sim NID(\mathbf{0}, V_{e\gamma}), \quad \mathbf{v}_\delta(k) \sim NID(\mathbf{0}, \delta^2 I)$$

where $\mathbf{s}_m(k)$ is a column vector of $s_{ij}(k)$s at the sites of $\Omega_b - \Omega_e$. By eliminating the coefficient on the left side of equation (3.25) as in Section 2.3, we obtain

$$\mathbf{s}_e(k) = \left[-S_e V_{e\gamma}^{-1}, \ldots, -S_e V_{e\gamma}^{-1}\right] \begin{pmatrix} \mathbf{s}_e(k-1) \\ \vdots \\ \mathbf{s}_e(k-11) \end{pmatrix} + \mathbf{v}_{se}(k)$$

$$\mathbf{v}_{se}(k) \sim NID(\mathbf{0}, S_e), \quad S_e = \{V_{e\gamma}^{-1} + \delta^{-2} D_e'(I - D_m D_m^+) D_e\}^{-1} \tag{3.26}$$

Using the equations (3.18), (3.22) and (3.26), we give the state-space model for space-time seasonal adjustment as:

$$\mathbf{y}_e(k) = F_e \mathbf{x}_e(k) + \mathbf{u}_e(k), \quad \mathbf{u}_e(k) \sim NID(\mathbf{0}, U_e) \tag{3.27}$$

$$\mathbf{x}_e(k) = G_e \mathbf{x}_e(k-1) + \mathbf{v}_e(k), \quad \mathbf{v}_e(k) \sim NID(\mathbf{0}, H_e) \tag{3.28}$$

where $\mathbf{x}_e(k) = (\mathbf{t}_e'(k), \mathbf{t}_e'(k-1), \mathbf{s}_e'(k), \ldots, \mathbf{s}_e'(k-10))'$,

$$F_e = [I, \mathbf{0}, I, \mathbf{0}, \ldots, \mathbf{0}],$$

$$G_e = \begin{bmatrix} 2T_eV_{e\alpha}^{-1} & -T_eV_{e\alpha}^{-1} & & & & \\ I & & & & & \\ & & -S_eV_{e\gamma}^{-1} & \cdots & -S_eV_{e\gamma}^{-1} & -S_eV_{e\gamma}^{-1} \\ & & & \ddots & & \\ & & & & I & \end{bmatrix}$$

and

$$H_e = \begin{bmatrix} T_e & & & & \\ & \mathbf{0} & & & \\ & & S_e & & \\ & & & \mathbf{0} & \\ & & & & \ddots \\ & & & & & \mathbf{0} \end{bmatrix}$$

The interpolations to estimate $\mathbf{t}_m(k)$ and $\mathbf{s}_m(k)$ are given by

$$\hat{\mathbf{t}}_m(k) = -D_m^+ D_e \hat{\mathbf{t}}_e(k), \quad \hat{\mathbf{s}}_m(k) = -D_m^+ D_e \hat{\mathbf{s}}_e(k)$$

where $\hat{\mathbf{t}}_e(k)$ and $\hat{\mathbf{s}}_e(k)$ are estimates of $\mathbf{t}_e(k)$ and $\mathbf{s}_e(k)$ respectively.

3.2. Remarks on the model

3.2.1. Spatial Uniformity of Variances

In the early stage of the development, we assumed spatial uniformity for the variances in the equations (3.17), (3.19) and (3.23) as in the equations (3.3) and (3.4). This was because the amount of computation required in maximizing the likelihood depends on the number of unknown variance parameters, though we had known that variation of water quality in Tokyo bay is spatially non-stationary in variance too. However, this assumption often caused over-smoothing. Consequently, in the proposed model it is assumed that the variances depend on the sites. At the same time, we took measures to reduce the amount of computation.

3.2.2. Modification of the Model

The proposed model is based on Kitagawa-Gersh's model and the DTPSM. However, these models are not unique, and there are several alternative models (see, for example, Ozaki and Thomson (1994) for time-series models, and Cressie (1991) for spatial models). It is easy to modify the proposed model based on such alternative models, if the generalized inverse technique mentioned in the previous sections is used.

3.2.3. Outlier

Together with missing values, the Tokyo Bay data involve outliers caused by irregular events such as typhoons, red water, etc. In the test of the proposed model, outliers caused unacceptable over-fitting even for the data to which time-series seasonal adjustment methods could be applied successfully. As a result, we decided to detect outliers, and treat detected ones as missing values. To detect outliers, we use Kitagawa-Gersh's time-series model.

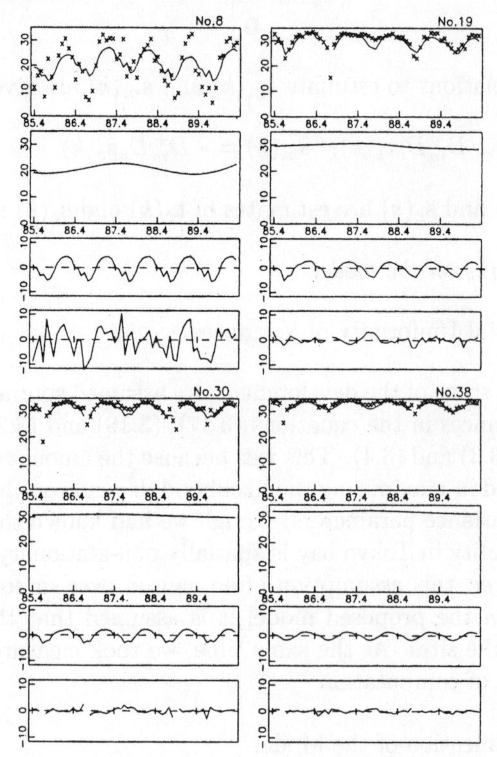

Figure 3.6: Time-series Plots of the Estimated Results at four Monitoring Points, No. 8, 19, 30 and 38.

3.3 *Application*

In this section, we show the results of the application of the proposed model to the data of salinity in the surface layer in Tokyo Bay measured at the 41 monitoring points every month from April 1985 to March 1990. Salinity is defined by the mixture ratio of sea water and fresh water coming in from the ocean and from the rivers respectively. The inflow of fresh water increases and decreases in summer and in winter respectively because of the weather conditions. Additionally, water exchange between the surface and bottom layers becomes inactive and active in summer and in winter respectively, because the specific gravity of water depends on both salinity and temperature, and because heat exchange between the atmosphere and the water occurs only in the surface layer. Consequently, seasonal changes of salinity in the surface layer are very clear. Using this data set, we checked the performance of the proposed model.

Figure 3.6 shows time-series plots of the estimated results at four monitoring points, No. 8, 19, 30 and 38 (see Figure 3.1 for their locations). The results at each monitoring point are shown by four plots. The top plot shows the data (cross) and the sum of the estimates of the trend and seasonal components (line). The remaining plots show the estimates of the trend, seasonal and residual components respectively. Lines in the top and bottom plots are broken at detected outliers. The horizontal and vertical axes of each plot indicate time (month) and salinity respectively.

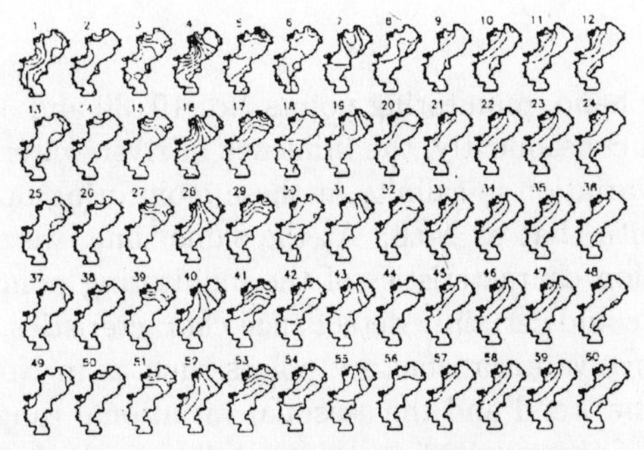

Figure 3.7: Contour Maps of the Estimated Trend of Salinity in the Surface Layer in Tokyo Bay Every Month from April 1985 to March 1990. Contour Lines are Drawn Every 1 Point Using a Thin Line (< 30), a Thick Line ($= 30$) and a Thin Line with Dots (> 30).

Monitoring point No. 8 is located near the mouth of the river. Consequently, salinity varies violently as is seen in the top plot. The proposed model could not follow such variation sufficiently. As may be seen from the bottom plot, the variance of the residual is very large and serial correlation remains in the sequence of the residuals. These are almost due to the fact that the maximum likelihood estimate of δ in equation (3.24) was very small, probably because the length of the sequence is too short. However, even from such data, the proposed model could estimate a specific trend as in the second plot. Although we cannot say exactly, because the drainage area of the river is considerably wide, this trend almost corresponds to a long-range variation in the precipitation.

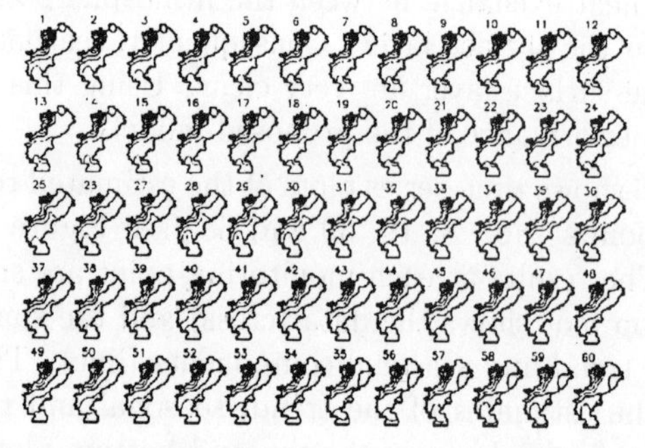

Figure 3.8: Contour Maps of the Estimated Seasonal Components of Salinity in the Surface Layer in Tokyo Bay Every Month from April 1985 to March 1990. Contour Lines are Drawn Every 1 Point Using a Thin Line (< 0), a Thick Line ($= 0$) and a Thin Line with Dots (> 0).

On the other hand, monitoring points No. 19, 30 and 38 are located offshore, and consequently, the influence of river water is rather indirect. The variation in salinity at these monitoring points is gentle comparing with that at No.8. At the same time, we notice that it reflects location characteristics of the monitoring points. The proposed model could certainly detect such characteristics, even though the results involve unsatisfactory points such as in No. 8. In fact, the level of the trend and the seasonal variational range depend on the position of the monitoring point relative to the big rivers, Edo, Ara and Tama. That the influence of river water can be evaluated is important because most of pollutants come into the bay from the rivers. Contour maps of the estimates of the trend and seasonal components respectively are shown in Figures 3.7 and 3.8. From these figures, we have inferred that the constant stream in the surface layer goes south along the west coast of the bay. This was not common before, because the stream in the surface layer in the bay is strongly affected by the tide and the wind. Additionally, it can be noted that the total amount of every measured quantity in the bay can be estimated by using the proposed model. The total amount is currently a valued environmental standard in Tokyo Bay.

REFERENCES

Akaike, H. (1980), 'Likelihood and the Bayes Procedure', *Trabajos de Estatistica*, 31, pp. 143-66.

Box, G.E.P. and G.M. Jenkins (1970), *Time Series Analysis, Forecasting and Control,* San Francisco: Holden-Day.

Cressie, N. (1991), *Statistics for Spatial Data,* New York: John Wiley.

Harvey, A.C. (1989), *Forecasting Structural Time Series Models and the Kalman Filter*, Cambridge: Cambridge University Press.

Hillmer, S.C. and G.C. Tiao (1982), 'An ARIMA-model-based Approach to Seasonal Adjustment', *Journal of the American Statistical Association*, 77, pp. 63-70.

Kashiwagi, N. (1997), 'A Study on Seasonal Adjustment by State-space Approach; an Application to Tokyo Bay Data' (in Japanese), *Proceedings of the Institute of Statistical Mathematics*, 45, pp. 329-42.

Kashiwagi, N. (1996), 'A State-space Approach to Polygonal Line Regression', *Annals of the Institute of Statistical Mathematics*, 48, pp. 215-28.

Kashiwagi, N. (1993), 'On Use of the Kalman Filter for Spatial Smoothing', *Annals of the Institute of Statistical Mathematics*, 45, pp. 21-34.

Kashiwagi, N. and T. Yanagimoto (1992), 'Smoothing Serial Count Data through a State-space Model', *Biometrics*, 48, pp. 1187-94.

Kitagawa, G. (1998), 'A Self-organizing State-space Model', *Journal of the American Statistical Association*, 93, pp. 1203-15.

Kitagawa, G. (1987), 'Non-Gaussian State-space Modeling of Nonstationary Time Series (with discussion)', *Journal of the American Statistical Association*, 82, pp. 1032-63.

Kitagawa, G. and W. Gersh (1984), 'A Smoothness Priors-state Space Modeling of Time Series with Trend and Seasonality', *Journal of the American Statistical Association*, 79, pp. 378-89.

Ninomiya, K., N. Kashiwagi and H. Ando (1996a), 'Seasonal Characteristics of Spatial Distributions of Water Temperature and Salinity in Tokyo Bay' (in Japanese), *Journal of Japan Society on Water Environment*, 19, pp. 480-90.

Ninomiya, K., N. Kashiwagi and H. Ando (1996b), 'Seasonal Characteristics of Spatial Distributions of COD and DO in Tokyo Bay' (in Japanese), *Journal of Japan Society on Water Environment*, 19, pp. 741-8.

Ninomiya, K., N. Kashiwagi, H. Ando and H. Ogura (1997), 'Seasonal Characteristics of Spatial Distributions of Dissolved Inorganic Nitrogen and Phosphorus in Tokyo Bay' (in Japanese), *Journal of Japan Society on Water Environment*, 20, pp. 457-67.

Ozaki, T. and P.J. Thomson (1994), 'A Dynamical Systems Approach to X-11 Type Seasonal Adjustment', *Research Memorandum*, No. 498, Tokyo: The Institute of Statistical Mathematics.

Shiskin, J., A.H. Young and J.C. Musgrave (1967), 'The X-11 Variant of the Census Method II Seasonal Adjustment Program', *Technical Paper 15*, Washington, D.C.: Bureau of the Census.

4

Statistical Issues in Environmental Evaluations

Bimal K. Sinha • Xiaoming Li • Barry Nussbaum

In this paper we discuss a few statistical problems relevant to environmental evaluations and review their solutions. The emphasis is on a clear description of the problems rather than on the solutions themselves. The problems reviewed here relate to assessing Superfund clean-up methods, checking gasoline quality via Reid Vapor Pressure, and examining water clarity in a lake.

1. Introduction

The statistical problems arising in the US Environmental Protection Agency (EPA) vary in their complexity and diversity. Since the mission of the EPA is to protect human health and to safeguard the natural environment–air, water, and land, in order to accomplish this with some scientific basis a good deal of environmental data must be analyzed. Emphasizing the importance of information in environmental decision-making, the EPA established an Office of Environmental Information (OEI) in 1999 by making appropriate changes in its former Center for Environmental Information and Statistics (CEIS). This office sees itself as an innovative centre of excellence that advances the creation, management, and use of information as a strategic resource to enhance public health and environmental protection.

In this paper we discuss three problems which demonstrate the spectrum of activity within OEI (and the former CEIS). They include

---

Our sincere thanks are due to J.K. Ghosh for his encouragement and helpful remarks.

a problem in Superfund clean-up, a gasoline quality assessment and an evaluation of water clarity. The emphasis will be on problem indentification and description rather than on the novelty of the solutions. All technical details are kept at a minimum.

The organization of the paper is as follows. In Section 2 we discuss the evaluation of superfund clean-up standards. In Section 3 we consider the problem of checking gasoline quality based on Reid Vapor Pressure (RVP). An analysis of Hillsdale Lake data is treated in Section 4. In Sections 3 and 4, we have used real data sets to demonstrate the methods.

2. Evaluation of Superfund Clean-up Standards

Before environmental responsibility became an everyday concern, people were not aware of how dumping chemical wastes might affect public health. Therefore, thousands of parcels of land became uncontrolled or abandoned hazardous waste sites. Citizens' concern over the extent to this problem led the US Congress to establish the Superfund programme in 1980 to locate, investigate and clean up the worst sites nationwide. In 1986, Superfund amendments were passed to make some important changes that would reflect EPA's experience in the first six years. One of these stressed the importance of permanent remedies and innovative treatment technologies in cleaning up the hazardous waste sites on the National Priorities List (NPL). Since these innovative remedial technologies are extremely expensive, a study of their performance after a certain amount of time is highly desirable. If found useful, these technologies can be encouraged to continue at the same site and be adopted at other sites too for the same remedial purposes. However, if such technologies are not performing satisfactorily, this should be determined as soon as possible so that suitable corrective measures can be taken.

Evaluating the attainment of a clean-up standard by a technology can be done in several ways. A common procedure is to study a *pre-remediation baseline sample* and an *interim sample* taken after a certain period of operation of the technology to find out if a predetermined percentage of the total contaminant has been removed. Although this is a familiar two-sample problem, as observed by Patil and Taillie (1990), the data in these studies are generally very skewed so that the use of standard normal theory is not tenable. Patil and

Taillie (1990) assumed independent gamma distributions with the same shape parameter and different scale parameters. They derived the likelihood ratio test (LRT) of the null hypothesis that a certain percentage of the contaminant mass has been reduced. Denoting the paired observations by $(X_1, Y_1), \ldots, (X_n, Y_n)$ corresponding to n locations within a superfund site, where X_is represent pre-remediation or baseline values and Y_is represent post-remediation or clean-up values, it is assumed that X_is are independent of Y_is, and that

$$X_i \sim f(x \mid \theta_1, \kappa) = \frac{e^{-x/\theta_1} x^{\kappa-1}}{\Gamma(\kappa)\theta_1^{\kappa}}, \theta_1 > 0, \kappa > 0 \tag{4.1}$$

and

$$Y_i \sim f(y \mid \theta_2, \kappa) = \frac{e^{-x/\theta_2} x^{\kappa-1}}{\Gamma(\kappa)\theta_2^{\kappa}}, \theta_2 > 0, \kappa > 0 \tag{4.2}$$

The null hypothesis H_0 that cleaning has reduced at least $100(1-\tau)$ percent of the contamination can then be expressed as

$$H_0 : \theta_2 \leq \tau\theta_1 \tag{4.3}$$

Authors O'Brien *et al.* (1991), using this assumption, provided the uniformly most powerful unbiased test of the null hypothesis H_0. Sinha and Sinha (1995) questioned the assumption of independence of the two samples (X:*before* clean-up and Y:*after* clean-up) on the ground that the very purpose of setting up a cleaning machinery is to reduce the level of contamination with the implication that the amount Y of contamination at a given location of the site after clean-up ought to be less than the amount X of contamination at the same location before clean-up. Thus, the two variables X and Y cannot be independent and in fact satisfy $0 < Y < X < \infty$. They proposed a class of bivariate exponential models, which are generalizations of Marshall and Olkin (1967)'s models and Sarkar (1987)'s models and discussed several tests of the same null hypothesis. As an example, a joint density of (X, Y) advocated by Sinha and Sinha (1995) is given by

$$
\begin{aligned}
f(x,y) &= \frac{f_1(x)f_2(y)}{F_2(y)-F_1(y)}\times \\
&\quad \exp\left[-\int_y^x \frac{dF_2(t)}{F_2(t)-F_1(t)}\right], \\
&\quad 0<y<x<\infty
\end{aligned} \tag{4.4}
$$

where $f_1(x)$ and $f_2(y)$ represent the marginal densities of X and Y, respectively. In particular, when $X \sim exp(\theta_1)$ and $Y \sim exp(\theta_2)$ with $\theta_1 = E(X) > E(Y) = \theta_2$, the joint density assumes the form

$$
\begin{aligned}
f_{\theta_1,\theta_2}(x,y) &= \frac{e^{-x\theta_1^{-1}-y\theta_2^{-1}}}{\theta_1\theta_2\left(e^{-y\theta_1^{-1}}-e^{-y\theta_2^{-1}}\right)} \\
&\quad \left[\frac{1-e^{-y(\theta_2^{-1}-\theta_1^{-1})}}{1-e^{-x(\theta_2^{-1}-\theta_1^{-1})}}\right]^{\theta_2^{-1}(\theta_2^{-1}-\theta_1^{-1})^{-1}}
\end{aligned} \tag{4.5}
$$

Sinha and Sinha (1995) discussed some properties of this joint distribution as also several tests of H_0 under this model.

We have discussed the parametric models so far. From a non-parametric point of view, Gilbert and Simpson (1990) proposed the Wilcoxon rank sum test and the quantile test to assess the attainment of site-specific standards at remediated Superfund sites. Odom and Sinha (1995) proposed several semi-parametric tests using the assumption that the distribution of contaminants from clean-up sites can be modelled as a non-parametric mixture of the distribution of background units and its *shifted* version, the amount of shift representing the magnitude of *dirts* clean-up. They discussed several large sample tests for the shift parameter as well as the mixing proportion under the following three models.

$$
\begin{aligned}
Model\ I: G(x) &= (1-\epsilon)F(x) \\
&\quad +\epsilon\, F(x-\theta), \theta \le 0,\ 0<\epsilon<1
\end{aligned} \tag{4.6}
$$

$$
\begin{aligned}
Model\ II: G(x) &= (1-\epsilon)F(x) \\
&\quad +\epsilon\, F(x/\theta),\ 0<\theta\le 1, \\
&\quad 0<\epsilon<1
\end{aligned} \tag{4.7}
$$

$$\begin{aligned} Model\ III:\ G(x) &= (1-\epsilon)F(x) \\ &+\epsilon F(\frac{x-\theta}{\beta}),\ \theta \le 0, \\ &0<\beta\le 1,\ 0<\epsilon<1 \end{aligned} \tag{4.8}$$

Here $F(x)$ is the cumulative distribution function (CDF) of background contamination, $G(x)$ is the CDF of clean-up units (after clean-up), θ, β, ϵ are unknown, and $F(x)$ is also assumed to be unknown. Models I and III assume that x spans over the entire real line and are thus appropriate in situations when deviations from contaminant from a given standard are measured, while x is assumed non-negative in Model II and can therefore represent the amount of contamination itself. The general spirit in these models is that the use of the technology removes only a part (ϵ) of the level of contamination, so that the distribution of the clean-up units results in a mixture of the distribution of background units with its shift. The parameter θ in models I and II and the parameters (θ, β) in model III represent precisely the nature of shift. Based on independent data sets $(X_1, \ldots, X_n) \sim iidF(x)$ and $(Y_1, \ldots, Y_m) \sim iidG(x)$ collected, respectively, before and after clean-up at a given site, Odom and Sinha (1995) discussed large sample tests of the hypotheses

$$H_{0I}: \theta = 0\ vs\ H_{1I}: \theta < 0 \tag{4.9}$$

$$H_{0II}: \theta = 1\ vs\ H_{1II}: \theta < 1 \tag{4.10}$$

$$H_{0III}: (\theta, \beta) = (0,1)\ vs\ H_{1III}: (\theta, \beta) \ne (0,1) \tag{4.11}$$

in models I, II and III respectively.

The above null hypotheses, if accepted, represent the situation that clean-up is not effective. Odom and Sinha (1995) also treated the case of paired data $\{(X_i, Y_i), i = 1, \ldots, n\}$ and also cases at the laboratory level which provide measurements.

3. Statistical Assessment of Gasoline Quality

The statistical model which is relevant here corresponds to a bivariate population of two variables (Y, Z) with the mean vector as (μ, μ) and a general dispersion matrix $\sum$ with the variances as σ^2, η^2 and correlation ρ. Suppose in addition to paired data $(y_i, z_i), i = 1, \ldots, m$ on (Y, Z), we also have data $x_i, i = 1, \ldots, n$

on one of the marginals, say, Y. Under this scenario, our goal is to make a meaningful inference about the common mean μ.

A novel application of the above formulation arises in the context of the EPA's attempt to evaluate the gasoline quality based on what is known as *Reid Vapor Pressure* (RVP). A major component of current efforts to reduce air pollution in large cities is the use of reformulated gasoline. The most effective gasoline control is to reduce the volatility of the gasoline as measured by RVP. To ensure that gas stations distribute gasoline that complies with clear air regulations, surveys of reformulated gasoline are performed regularly. Typically, samples of gasoline are taken from various pumps and RVPs are measured in two ways: at the field level, which provides cheap and quick measurements and also at the laboratory level which provides measurements of presumably higher precision. Since laboratory measurements are usually much more expensive than field measurements because of the special packaging used to ship a gas sample from a pump to a laboratory, most of the time only field measurements are taken. Some times data are collected at both levels. If the average RVP in a supply of gasoline is denoted by μ, we can assume that the field data and the laboratory data share the common mean μ with possibly unequal variances σ^2 and η^2. Moreover, when paired data (field, laboratory), are observed, there is a natural dependency between them with covariance $\sigma\eta\rho$ while field-only data are distributed marginally with mean μ and variance σ^2. Assuming a bivariate normal or log-normal distribution of (field, laboratory) data, the problem of drawing suitable inferences about μ has been addressed by Yu *et al.* (1999). Nussbaum and Sinha (1995) and Yu *et al.* (1999) discussed this problem using a ***ranked set sampling*** approach. Moreover, Li (2000) has provided a Bayesian solution to this problem. We provide below a review of some of the solutions on the basis of a simple random sample (both paired and marginal).

Denote the 'field-only' data by $x_i, i = 1, \ldots, n$, and the paired 'field-laboratory' data by $(y_j, z_j), j = 1, \ldots, m$. Then, under our assumption

$$E(x_i) = \mu,\ var(x_i) = \sigma^2 \text{ for } i = 1, \ldots, n,$$

$$E(y_j) = E(z_j) = \mu,\ var(y_j) = \sigma^2,\ var(z_j) = \eta^2,$$
$$cov(y_j, z_j) = \sigma\eta\rho, j = 1, \ldots, m.$$

The dispersion matrix $\sum$ of (y_j, z_j) can be written as

$$\sum = \begin{bmatrix} \sigma^2 & \xi \\ \xi & \eta^2 \end{bmatrix}, \quad \xi = \sigma\eta\rho$$

Inference for μ and $\sum$ are based on $(\bar{x}, \bar{y}, \bar{z}, s_x^2, s_z^2, s_{yz})$ where $\bar{x}, \bar{y}$ and $\bar{z}$ are sample means and

$$s_x^2 = \frac{\sum_1^n (x_i - \bar{x})^2}{(n-1)}, \quad s_y^2 = \frac{\sum_1^m (y_i - \bar{y})^2}{(m-1)},$$

$$s_z^2 = \frac{\sum_1^m (z_i - \bar{z})^2}{(m-1)}, \quad s_{yz} = \frac{\sum_1^m (y_i - \bar{y})(z_i - \bar{z})}{(m-1)}$$

We first estimate μ for a given $\sum$ by

$$\hat{\mu} = \frac{\frac{n}{\sigma^2}\bar{x} + m\left[\frac{\eta^2-\xi}{\sigma^2\eta^2-\xi^2}\bar{y} + \frac{\sigma^2-\xi}{\sigma^2\eta^2-\xi^2}\bar{z}\right]}{\frac{n}{\sigma^2} + m\frac{\sigma^2+\eta^2-2\xi}{\sigma^2\eta^2-\xi^2}} \tag{4.12}$$

which is the maximum likelihood estimate (MLE) of μ under normality assumption, and is always unbiased with variance

$$var(\hat{\mu}) = \frac{1}{\frac{n}{\sigma^2} + m\frac{\sigma^2+\eta^2-2\xi}{\sigma^2\eta^2-\xi}} \tag{4.13}$$

A convenient choice for estimation of $\sum$ is given by

$$\hat{\sigma}_1^2 = \frac{(n-1)s_x^2 + (m-1)s_y^2}{n+m-2}, \hat{\eta}_1^2 = s_x^2, \hat{\xi}_1 = s_{yz} \tag{4.14}$$

However, the resulting $\hat{\sum}_1$ is not always a non-negative definite (*nnd*). Yu *et al.* (1999) suggested the following estimate of $\sum$ (known as REML) which is always *nnd*.

$$\hat{\sigma}_2^2 = \frac{(n-1)s_x^2 + (m-1)s_y^2}{n+m-2} \tag{4.15}$$

$$\hat{\eta}_2^2 = \sigma_2^2 \frac{s_{yz}^2}{s_y^4} + \frac{s_y^2 s_z^2 - s_{yz}^2}{s_y^2} \tag{4.16}$$

$$\hat{\xi}_2 = \hat{\sigma}_2^2 \frac{s_{yz}}{s_y^2} \tag{4.17}$$

Hence, a suitable estimate of μ when the value of $\sum$ is unknown is obtained by plugging the equations (4.15) to (4.17) into equation (4.12) which results in

$$\widetilde{\mu} = \frac{\frac{n}{\hat{\sigma}^2}\bar{x} + m\left[\frac{\hat{\eta}^2 - \xi}{\hat{\sigma}^2\hat{\eta}^2 - \xi^2} + \frac{\hat{\sigma}^2 - \xi}{\hat{\sigma}^2\hat{\eta}^2 - \xi^2}\bar{z}\right]}{\frac{n}{\hat{\sigma}^2} + m\frac{\hat{\sigma}^2 + \hat{\eta}^2 - 2\xi}{\hat{\sigma}^2\hat{\eta}^2 - \xi^2}} \tag{4.18}$$

An exact expression for $Var(\widetilde{\mu})$ is usually very difficult to obtain. However, for the purpose of inferences, what is really needed is an estimate of $Var(\widetilde{\mu})$ and several estimates are suggested in Yu *et al.* (1999). The simplest estimator of $Var(\widetilde{\mu})$ is obtained by plugging the estimator $\hat{\sum}$ of $\sum$ in $var(\hat{\mu})$ given by the equation (4.13) which leads to

$$var(\widetilde{\mu}) = \frac{1}{\frac{n}{\hat{\sigma}^2} + m\frac{\hat{\sigma}^2 + \hat{\eta}^2 - 2\xi}{\hat{\sigma}^2\hat{\eta}^2 - \xi^2}} \tag{4.19}$$

It has been pointed out by many investigators that the above method is likely to underestimate $Var(\widetilde{\mu})$.

We now discuss briefly the problem of testing the hypothesis

$$H_0 : \mu = \mu_0 \ vs. \ H_1 : \mu \neq \mu_0 \tag{4.20}$$

where μ_0 is a specified value, and of constructing a confidence interval (CI) for μ.

It is well-known that based on $(\bar{x}, s_x^2)$ alone, we can use the well-known single sample t-test to test equation (4.20). The test statistic and CI are given by

$$t_1 = \frac{\bar{x} - \mu_0}{\sqrt{s_x^2/n}} \tag{4.21}$$

$$\begin{aligned} 100(1-\alpha)\%C.I. &= \left\{\mu_0 : |t_1| \leq t_{\frac{\alpha}{2},n-1}\right\} \\ &= \left[\bar{x} - t_{\frac{\alpha}{2}.n-1}\sqrt{\frac{s_x^2}{n}}, \bar{x} \right. \\ &\qquad \left. + t_{\frac{\alpha}{2}.n-1}\sqrt{\frac{s_x^2}{n}}\right] \end{aligned} \tag{4.22}$$

where t_1 has $n-1$ degrees of freedom.

To derive a test for μ based on the paired data (y, z), we use the following transformation.

Let $u = (y+z)/2$ and $v = (y-z)/2$. Then

$$\begin{pmatrix} u_j \\ v_j \end{pmatrix} \sim N\left\{\begin{pmatrix} \mu \\ 0 \end{pmatrix}, \frac{1}{4}\begin{bmatrix} \sigma^2+\eta^2+2\xi & \sigma^2-\eta^2 \\ \sigma^2-\eta^2 & \sigma^2+\eta^2-2\xi \end{bmatrix}\right\} \quad (4.23)$$

Let $\overline{u}, \overline{v}, s_u^2, s_v^2, s_{uv}$ be the usual sample means, sample variances and covariance based on the data (u_j, v_j), $j = 1, \ldots, n$. Then, it readily follows that

$$\begin{aligned} \overline{u} &= \frac{\overline{y}+\overline{z}}{2}, \overline{v} = \frac{\overline{y}-\overline{z}}{2}, s_u^2 = \frac{1}{4}(s_y^2 + 2s_{yz} + s_z^2) \\ s_v^2 &= \frac{1}{4}(s_y^2 - 2s_{yz} + s_z^2), s_{uv} = \frac{1}{4}(s_y^2 - s_z^2) \end{aligned}$$

Following Srivastava and Khatri (1979), the likelihood ratio test (LRT) statistic is given by

$$\mathcal{L} = \frac{1 + \frac{m}{m-1}\frac{(\overline{u}-\mu_0)^2 s_y^2 - 2(\overline{u}-\mu_0)\overline{v}s_{uv} + \overline{v}^2 s_u^2}{s_u^2 s_v^2 - s_{uv}^2}}{1 + \frac{m}{m-1}\frac{\overline{v}^2}{s_v^2}} \quad (4.24)$$

and

$$(m-2)(\mathcal{L}-1) \sim F_{1,m-2} \quad (4.25)$$

By converting the acceptance region of the LRT, we get the CI of μ as

$$\begin{aligned} 100(1-\alpha)\%C.I. &= \{\mu_0 : (m-2)(\mathcal{L}-1) \\ &\quad < F_{\alpha;1,m-2}\} \\ &= \left(\overline{u} - \frac{\overline{v}s_{uv}}{s_v^2} \right. \\ &\quad -\sqrt{hF_{\alpha;1,m-2}} \\ &\quad \overline{u} - \frac{\overline{v}s_{uv}}{s_v^2} \\ &\quad \left. +\sqrt{hF_{\alpha;1,m-2}}\right) \end{aligned} \quad (4.26)$$

where

$$\overline{h} = \frac{1}{m-2}\left(\frac{m-1}{m} + \frac{\overline{v}^2}{s_v^2}\right)\frac{(s_u^2 s_v^2 - s_{uv}^2)}{s_v^2}$$

It is now necessary to combine the above two tests meaningfully into a single test. Several methods to accomplish this are available in the literature. Let

$$\begin{aligned} t_2 &= \frac{\overline{u} - \mu_0 - \frac{\overline{v} s_{uv}}{s_v^2}}{\sqrt{\overline{h}}} \\ i.e., \ |t_2| &= \sqrt{(m-2)(\mathcal{L}-1)} \\ F_1 &= t_1^2, \ F_2 = (m-2)(\mathcal{L}-1) \end{aligned}$$

so that

$t_2 \sim t_{m-2}, \ F_1 \sim F_{1,n-1}, F_2 \sim F_{1,m-2}$

Define

$$P_1 = \int_{F_1}^{\infty} f_{1,n-1}(x)dx, \ P_2 = \int_{F_2}^{\infty} f_{1,m-2}(x)dx$$

Cohen and Sackrowitz (1984) suggested the use of

$M_t = \max_{1 \leq i \leq 2} \{|t_i|\}$ as a test statistic for testing hypotheses about μ. In our case, if a cut-off point $c_{\alpha/2}$ satisfies the condition

$$1 - \alpha = P[M_t \leq c_{\alpha/2}]$$

then we reject H_0 when $M_t > c_{\alpha/2}$ and an exact confidence interval for μ with confidence level $1-\alpha$ is given by $[LB_{CS}, UB_{CS}]$ where

$$LB_{CS} = \max\left\{\overline{x} - c_{\alpha/2}\sqrt{\frac{s_x^2}{n}}, \overline{u} - \frac{\overline{v}_{uv}}{s_v^2} - c_{\alpha/2}\sqrt{\overline{h}}\right\} \tag{4.27}$$

$$UB_{CS} = \min\left\{\overline{x} + c_{\alpha/2}\sqrt{\frac{s_x^2}{n}}, \overline{u} - \frac{\overline{v}_{uv}}{s_v^2} + c_{\alpha/2}\sqrt{\overline{h}}\right\} \tag{4.28}$$

Determination of the cut-off point $c_{\alpha/2}$ is not easy, and simulation may be necessary.

Fairweather (1972) suggested using a weighted linear combination of the t_is namely,

$$W_t = \sum_{i=1}^{2} \gamma_i t_i, \ \gamma_i = \frac{(var(t_i))^{-1}}{\sum_{j=1}^{2}(var(t_j))^{-1}} \tag{4.29}$$

In our case,

$$\gamma_1 = \frac{(n-3)(m-2)}{(n-3)(m-2)+(n-1)(m-4)}$$

and $\gamma_2 = 1 - \gamma_1$.

If $b_{\alpha/2}$ denotes the cut-off point of the distribution of W_t, sattisfying the equation

$$1 - \alpha = P[|W_t| \le b_{\alpha/2}]$$

then the confidence interval for μ is obtained as $[LB_{FW}, UB_{FW}]$ where

$$LB_{FW} = \frac{\frac{\gamma_1 \overline{x}}{\sqrt{s_x^2/n}} + \frac{\gamma_2}{\sqrt{h}}(\overline{u} - \overline{v}\frac{s_{uv}}{s_v^2}) - b_{\alpha/2}}{\frac{\gamma_1}{\sqrt{s_x^2/n}} + \frac{\gamma_2}{\sqrt{h}}} \tag{4.30}$$

$$UB_{FW} = \frac{\frac{\gamma_1 \overline{x}}{\sqrt{s_x^2/n}} + \frac{\gamma_2}{\sqrt{h}}(\overline{u} - \overline{v}\frac{s_{uv}}{s_v^2}) + b_{\alpha/2}}{\frac{\gamma_1}{\sqrt{s_x^2/n}} + \frac{\gamma_2}{\sqrt{h}}} \tag{4.31}$$

The cut-off point $b_{\alpha/2}$ is usually obtained by simulation. Jordan and Krishnamoorthy (1996) have given some cut-off points for $\alpha =$ 0.05 and 0.01.

Jordan and Krishnamoorthy (1996) suggested using a linear combination of the F_is such as

$$W_f = \sum_{i=1}^{2} \gamma_i F_i$$

as a test statistic, where positive weights γ_is are inversely proportional to $var(F_i)$. Hence, if we can obtain a_α such that

$P[W_f \le a_\alpha] = 1 - \alpha$, then, after simplification, an exact confidence interval for μ with confidence level $1 - \alpha$ is given by $[LB_{JK}, UB_{JK}]$ where

$$LB_{JK} = \frac{A - B}{C} \tag{4.32}$$

$$UB_{JK} = \frac{A + B}{C} \tag{4.33}$$

where

$$A = \frac{\gamma_1 \overline{x}}{s_x^2/n} + \frac{\gamma_2}{\overline{h}}(\overline{u} - \overline{v}\frac{s_{uv}}{s_v^2})$$

$$B = \sqrt{a_\alpha(\frac{\gamma_1}{s_x^2/n} + \frac{\gamma_2}{\overline{h}}) - \frac{\gamma_1}{s_x^2/n}\frac{\gamma_2}{\overline{h}}(\overline{x} - \overline{u} + \overline{v}\frac{s_{uv}}{s_v^2})^2}$$

$$C = \frac{\gamma_1}{s_x^2/n} + \frac{\gamma_2}{\overline{h}}$$

Here $\gamma_2 = 1 - \gamma_1$ and

$$\begin{aligned} \gamma_1 &= (n-3)^2(n-5)(m-2)^2(m-3)/\left[(n-3)^2 \right. \\ &\quad \times (n-5)(m-2)^2(m-3) + (n-1)^2 \\ &\quad \left. \times (n-2)(m-4)^2(m-6)\right] \end{aligned} \tag{4.34}$$

Again, a_α is usually obtained by simulation. Jordan and Krishnamoorthy (1996) gave some cut-off points for $\alpha = 0.05$ and 0.01.

The CIs for μ may also be obtained by combining P_1 and P_2 suitably. Tippett (1931)s method yields the same result as equations (4.27) and (4.28). Fisher (1932) suggested using $-2[\log(P_1) + \log(P_2)]$ as a test statistic for testing equation (4.20) and rejecting H_0 if

$$-2[\log(P_1) + \log(P_2)] > \chi^2_{4,\alpha}$$

Converting the acceptance region, we get the $100(1-\alpha)$ percent CI as

$$\{\mu_0 : -2[\log(P_1) + \log(P_2)] \leq \chi^2_{4,\alpha}\} \tag{4.35}$$

Similarly, the CI can be obtained by the inverse normal method (Stouffer *et al.* 1949) thus:

$$\left\{ \mu_0 : \frac{\sum_{i=1}^{2} \Phi^{-1}(P_i)}{\sqrt{2}} \geq -z_\alpha \right\} \tag{4.36}$$

where Φ denotes the cumulative probability function of the standard normal distribution. Finally, the CI based on the logit method (George 1977) can be obtained as

$$\left\{ \mu_0 : \left[-\sum_{i=1}^{2} \log \left(\frac{P_i}{1 - P_i} \right) \right] \left[\frac{3}{2\pi^2} \right]^{1/2} \leq z_\alpha \right\} \tag{4.37}$$

Tests based on the inverse normal and logit methods can be obtained similarly. It should be pointed out that some methods may not always produce genuine intervals. For details, we refer to Yu *et al.* (1999).

We now return to the application of the preceding theory to gasoline data. A random sample of size 55 was drawn from field-data and an independent paired sample of size 15 was drawn from (field-laboratory) data. The data set is presented in Table 4.1.

A direct computation yields the following basic statistics:

$$\overline{x} = 7.6247, s_x^2 = 0.1966, \overline{y} = 8.2489, s_y^2 = 0.3041$$

$$\overline{z} = 8.2760, s_z^2 = 0.2252, s_{yz} = 0.2558, r_{yz} = 0.9773.$$

The REML estimate of Σ turns out to be

$$\widehat{\sigma}^2 = 0.2187, \ \widehat{\eta}^2 = 0.1648, \ \widehat{\xi} = 0.1839.$$

Plugging the above values into equations (4.18) and (4.19), we get an estimate of μ as $\widetilde{\mu} = 7.8281$ with its estimated standard error as 0.0556.

On the other hand, the 95 percent confidence interval for μ based on only the field data comes out as (7.5049, 7.7446), that based on the paired (field, laboratory) data as (8.0771, 8.5684), that based on Fairweather's method as (7.7328, 7.9578), and finally the one based on the normal approximation $\widetilde{\mu} \pm 1.96 s.e.$ comes out as (7.7191, 7.9370). These results are summarized in Table 4.2. Unfortunately, it turns out that a confidence interval for μ for all the other procedures does *not* exist.

Table 4.1

Field and Laboratory Data on RVP for Regular Gasoline *

x	7.51	7.41	7.63	7.43	7.80	7.16	7.47	7.45	8.80	7.74
	7.01	7.45	7.37	7.56	8.09	8.51	7.43	7.38	8.02	7.93
	7.52	7.44	7.22	7.57	7.64	8.69	7.40	7.38	8.03	7.96
	6.53	7.32	7.47	7.53	7.98	8.46	7.41	7.47	8.01	7.86
	7.01	7.31	7.54	7.35	7.88	7.37	7.37	7.19	7.69	7.70
	7.31	7.59	7.28	8.90	7.83					
(y, z)	(7.56, 7.63)	(9.28, 9.14)	(9.28, 9.15)	(8.60, 8.52)	(8.56, 8.48)					
	(8.64, 8.63)	(8.70, 8.62)	(7.83, 7.90)	(7.88, 7.89)	(7.99, 8.32)					
	(8.03, 8.28)	(7.86, 7.85)	(7.86, 7.86)	(7.83, 7.92)	(7.83, 7.95)					

* Data Source: Private communication

Table 4.2

Point Estimation and CI for RVP of Regular Gasoline

Parameter	Result
$\widetilde{\mu}$	7.8281
S.E. of $\widetilde{\mu}$	0.0556
CI based on F only	(7.5049, 7.7446)
CI based on Pair	(8.0771, 8.5684)
CI based on Fairweather	(7.7328, 7.9578)
CI based on $\widetilde{\mu} \pm 1.96 s.e.$	(7.7191, 7.9370)

Finally, we indicate a Bayesian analysis which is carried out in Li (2000) under two scenarios: marginal data X has mean μ and variance σ^2 as above (scenario I) and marginal data X has mean $\mu+\delta$ and variance σ^2 (scenario II). Here, δ represents a possible negative bias in the X measurements. Under a non-informative prior given by

$$\pi(\mu, \sigma, \eta, \rho) = \frac{1}{\sigma\eta\sqrt{1-\rho^2}} \tag{4.38}$$

under scenario I, and by

$$\pi(\mu, \delta, \sigma, \eta, \rho) = \frac{1}{\sigma\eta\sqrt{1-\rho^2}} \tag{4.39}$$

under scenario II, where $\sigma > \eta > 0$, the respective full conditionals of parameters are obtained as

$$\pi(\mu|\mathbf{x},\mathbf{y},\mathbf{z}) \propto \prod_{i=1}^{n} \exp\left\{-\frac{(x_i-\mu)^2}{2\sigma^2}\right\} \prod_{j=1}^{m} \exp\left\{-\frac{1}{2(1-\rho^2)} \left(\left(\frac{y_j-\mu}{\sigma}\right)^2 - 2\rho\left(\frac{y_j-\mu}{\sigma}\right)\left(\frac{z_j-\mu}{\eta}\right) + \left(\frac{z_j-\mu}{\eta}\right)^2\right)\right\} \tag{4.40}$$

$$\pi(\sigma|\mathbf{x},\mathbf{y},\mathbf{z}) \propto \frac{1}{\sigma^{m+n+1}} \prod_{i=1}^{n} \exp\left\{-\frac{(x_i-\mu)^2}{2\sigma^2}\right\} \prod_{j=1}^{m} \exp\left\{-\frac{1}{2(1-\rho^2)} \left(\left(\frac{y_j-\mu}{\sigma}\right)^2 -2\rho\left(\frac{y_j-\mu}{\sigma}\right)\left(\frac{z_j-\mu}{\eta}\right) + \left(\frac{z_j-\mu}{\eta}\right)^2\right)\right\} \tag{4.41}$$

$$\pi(\eta|\mathbf{y},\mathbf{z}) \propto \frac{1}{\eta^{m+1}} \prod_{j=1}^{m} \exp\left\{-\frac{1}{2(1-\rho^2)} \left(\left(\frac{y_j-\mu}{\sigma}\right)^2 - 2\rho\left(\frac{y_j-\mu}{\sigma}\right)\left(\frac{z_j-\mu}{\eta}\right) + \left(\frac{z_j-\mu}{\eta}\right)^2\right)\right\} \tag{4.42}$$

$$\pi(\rho|\mathbf{y},\mathbf{z}) \propto \frac{1}{(1-\rho^2)^{\frac{m+1}{2}}} \prod_{j=1}^{m} \exp\left\{-\frac{1}{2(1-\rho^2)} \left(\left(\frac{y_j-\mu}{\sigma}\right)^2 - 2\rho\left(\frac{y_j-\mu}{\sigma}\right)\left(\frac{z_j-\mu}{\eta}\right) + \left(\frac{z_j-\mu}{\eta}\right)^2\right)\right\} \tag{4.43}$$

under scenario I. Similar results can be obtained under scenario II.

The full conditional distributions in equations (4.40) through (4.43) were used to obtain Markov Chain Monte Carlo samples from the joint posterior distribution μ, σ, η, ρ and δ (for scenario II). For the computations, many versions of Markov Chain Monte Carlo methods can be used. For a detailed description of these methods, see Besag and Green (1993) and the references therein. A natural choice is the Gibbs sampler (Geman and Geman, 1984). However, given the computational demands placed by sampling from the full conditionals, we used a Hastings algorithm (Hastings, 1970). Samples from the posterior distributions were generated by running the sampler from an initial random realization with a sequential visiting schedule and a burn-in period of 10,000 iterations. A sub-sample of size 500 - comprising every hundredth realization in the Markov Chain after burn-in - was stored and used to carry out the inference.

Table 4.3

Simulation Results for Scenarios I and II

Parameter	μ	δ	σ	η	ρ
Scenario I					
Actual α	3.5 %	*	4.5 %	4.5 %	3.0 %
Scenario II					
Actual α	4.5 %	3.5 %	5.0 %	6.5%	4.0%

* Not applicable for Scenario I

A total of 200 simulations were performed with $\mu = 7.80$, $\sigma = 0.60$, $\eta = 0.40$, $\rho = 0.97$, $n = 55$, $m = 15$ and $\alpha = 0.05$ for scenario I and $\mu = 8.30$, $\delta =$ -0.7, $\sigma = \sqrt{0.3}, \eta = \sqrt{0.22}, \rho = 0.97$, $n = 55$, $m = 15$ and $\alpha = 0.05$ for Scenario II. The results are shown in Table 4.3.

Both scenarios I and II are applied to the RVP data set and the results are summarized in Table 4.4 and Figure 4.1 and Table 4.5 and Figure 4.2 separately.

We conclude this section with the following observation. It can be seen that the estimates of σ^2 and η^2 bear out the expectation that the laboratories were doing a better job of measuring the RVP. It seems that even the means show a difference. There is a strong indication of the downward bias of the field estimates of RVP as supported in the second scenario. This can indeed be a matter of concern depending on what the EPA guidelines are regarding RVP. Since the lower the level of RVP the better, data analysis under the second scenario might lead to some gas stations being in violation of fair practice. In particular, a hypothetical EPA threshold value of 8.0 for mean

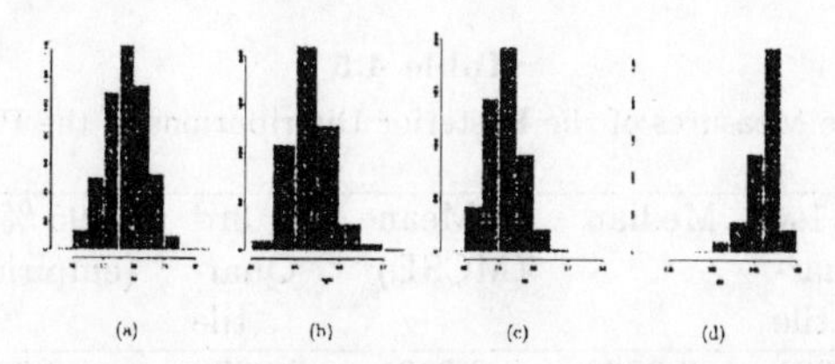

Figure 4.1: Histograms of Estimated Marginal Posterior Distributions of Parameters (a) μ, (b) σ, (c) η, (d) ρ

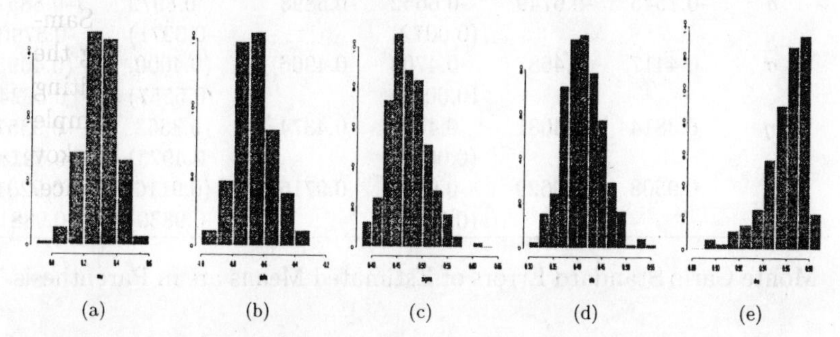

Figure 4.2: Histograms of Estimated Marginal Posterior Distributions of Parameters (a) μ, (b) δ, (c) σ, (d) η and (e) ρ

RVP yields different conclusions under the two scenarios. The EPA would tend to adopt the more stringent conclusion to safeguard the public.

Table 4.4

Descriptive Measures of the Posterior Distributions of the Parameters

Parameter	1st Quartile	Median	Mean (MCSE)	3rd Quartile	95% CI (empirical)	95% CI (shortest)
μ	7.7327	7.7759	7.7768 (0.0032)	7.8205	(7.6484, 7.9010)	(7.6434, 7.8931)
σ	0.5027	0.5305	0.5333 (0.0023)	0.5607	(0.4528, 0.6357)	(0.4514, 0.6317)
η	0.4785	0.5136	0.5153 (0.0025)	0.5468	(0.4232, 0.6155)	(0.4207, 0.6117)
ρ	0.9520	0.9643	0.9605 (0.0011)	0.9734	(0.9131, 0.9835)	(0.9248, 0.9869)

Monte Carlo Standard Errors of Estimated Means are in Parenthesis

Table 4.5

Descriptive Measures of the Posterior Distributions of the Parameters

Parameter	1st Quartile	Median	Mean (MCSE)	3rd Quartile	95 % CI (empirical)	95 % CI (shortest)
μ	8.2231	8.2960	8.8321 (0.0062)	8.3639	(8.0590, 8.4926)	(8.1040, 8.5107)
δ	-0.7545	-0.6749	-0.6652 (0.0072)	-0.5898	(-0.8979 -0.3971)	(-0.8857, -0.3790)
σ	0.4417	0.4687	0.4708 (0.0018)	0.4966	(0.4000, 0.5557)	(0.4095, 0.5624)
η	0.3814	0.4081	0.4104 (0.0020)	0.4374	(0.3363, 0.4923)	(0.3357, 0.4918)
ρ	0.9508	0.9629	0.9889 (0.0009)	0.9716	(0.9116, 0.9830)	(0.9201, 0.9881)

Monte Carlo Standard Errors of Estimated Means are in Parenthesis

4. Assessment of Water Clarity of Hillsdale Lake

Hillsdale Lake, a large federal reservoir located about 30 miles from the Kansas City metropolitan area, was authorized by Congress in 1954 as part of a comprehensive flood control plan for the Osage and Missouri River Basins. However, the lake has become a major recreational resource, receiving about 500,000 visitors annually, and is a significant source of drinking water for much of southern Johnson county and northern Miami county. The Hillsdale Water Quality Project Inc., initiated in 1993 and currently a non-profit organization, was developed to prevent degradation of Hillsdale Lake. Under a co-operative agreement with EPA (with funding provided under Section 104(b)(3) of the Clean Water Act), one of the major tasks is to conduct a survey of lake users to establish what level of water clarity is perceived as *good.* User opinion, as determined by this survey, will be a factor in establishing water quality standard for the lake. It will be incorporated subsequently into a sediment Total Maximum Daily Load for the lake and its contributing watershed.

The object of this section is to provide a brief statistical analysis of the data collected over the summer of 1999 from various lake users. Data were collected from various categories of lake uses: swimming, fishing, boating, skiing and water sports. On days when data were collected, a measure of *secchi* depth, denoted by d, was obtained. The response of the lake users in terms of how they perceived the water quality on that day were recorded in three categories: *poor, good* and *excellent.* Values of several covariates such as the number of people in a car, temperature, weather condition (sunny, cloudy/rainy) were also observed. We summarize the relevant data. We have ignored the various classifications of lake users and have performed the statistical analysis based on total users. Moreover, the number of lake users on 21 June, 1999 and 26 July 1999 are not available, and we have estimated them under the assumption of three people per car on the average.

Following the well-known *dose-response* models, we formulate our statistical model in a way which connects the response meaningfully in terms of percentage (P) of lake users in combined *good* and *excellent* categories (to total users) with the *secchi* depth d. It is expected that the larger the value of d, the larger will be the percentage P of *satisfied* lake users! There are two statistical models in the literature which are frequently used in this context: logit and probit. For brevity, we discuss only the logit model without covariates. For details, we refer to Li *et al.* (1999). Details of the derivation which follow below can be found in Agresti (1990), Berkson (1955), Shen (2000) and Sinha (1988).

4.1 The Logit Model (without covariates)

Here $L = ln(P/Q)$, where
$Q = 1 - P$, is modelled as

$$L = \alpha + \beta d \tag{4.44}$$

where α is the intercept and β is the slope. This is equivalent to assuming that P satisfies:

$$P = \frac{e^{\alpha+\beta d}}{1 + e^{\alpha+\beta d}} \tag{4.45}$$

Clearly, when $\beta > 0, P$ is an increasing function of d. In practice, the two parameters α and β will be unknown, and are estimated based on the data $\{p_i, d_i\}$, $i = 1, \ldots, k$ where $p_i = x_i/n_i$ is the sample percentage of *satisfied* lake users on the ith day (combining both good quality and excellent quality response) and d_i is the *secchi* depth measured on the ith day, and k is the total number of days

over the entire summer when data are collected. The two extremes when $p_i = 0$ or 1, which would imply
$\log(p_i/(1-p_i))$ approach $-\infty$ or ∞
are avoided by taking $p_i = 0.05$ or 0.95 respectively.

The weighted least squares estimates of $\theta = (\alpha, \beta)'$ are obtained by solving the linear equations:

$$\mathbf{A}\theta = \mathbf{z} \tag{4.46}$$

where

$$\mathbf{A} = \begin{bmatrix} \sum_{i=1}^{k} n_i p_i q_i & \sum_{i=1}^{k} d_i n_i p_i q_i \\ \sum_{i=1}^{k} d_i n_i p_i q_i & \sum_{i=1}^{k} d_i^2 n_i p_i q_i \end{bmatrix}, \mathbf{z} = \begin{bmatrix} \sum_{i=1}^{k} l_i n_i p_i q_i \\ \sum_{i=1}^{k} l_i d_i n_i p_i q_i \end{bmatrix}$$

and

$$l_i = \log(p_i/q_i)$$

Based on equation (4.13), the *secchi* depth d_0, which would guarantee a desired level of satisfaction P_0, is given by

$$d_0 = \frac{\log\left(\frac{P_0}{1-P_0}\right) - \alpha}{\beta} \tag{4.47}$$

Obviously, d_0 is unknown since α and β are so. Using the estimates of α and β, d_0 is estimated as

$$\widehat{d_0} = \frac{\log\left(\frac{P_0}{1-P_0}\right) - \widehat{\alpha}}{\widehat{\beta}} \tag{4.48}$$

with the estimated standard error of $\widehat{d_0}$ as

$$est.se(\widehat{d_0}) \approx \sqrt{\widehat{\phi}'\widehat{\Sigma}\widehat{\phi}} \tag{4.49}$$

Here, Σ stands for the covariance matrix of $\widehat{\theta}$ and its estimate $\widehat{\Sigma}$ is given by

$$\widehat{\Sigma}(\widehat{\theta}) = \begin{bmatrix} \sum_{i=1}^{k} n_i p_i q_i & \sum_{i=1}^{k} d_i n_i p_i q_i \\ \sum_{i=1}^{k} d_i n_i p_i q_i & \sum_{i=1}^{k} d_i^2 n_i p_i q_i \end{bmatrix}^{-1}.$$

Also

$$\widehat{\phi} = \begin{bmatrix} -1/\widehat{\beta} \\ \log\left(\dfrac{P_0}{1-P_0}\right) - \widehat{\alpha}\,\widehat{\beta}^2 \end{bmatrix}$$

The 95 percent confidence interval of the true *secchi* depth d_0 is then given by

$$[\widehat{d}_0 - 1.96\ est.se(\widehat{d}_0), \widehat{d}_0 + 1.96 est.se(\widehat{d}_0)]$$

It is also possible to derive valid inference for d_0 based on the logit model (with covariates) as well as probit models (with and without covariates). We omit the details (Li *et al.* 1999).

We now apply the method outlined above on the data displayed in Table 4.6. The results of our statistical analysis are summarized in Table 4.7. We have considered $P_0 = 80$ percent.

Based on the above analysis, we have prepared Table 4.8 which provides a 95 percent upper bound (d_{upper}). The interpretation is that if a water clarity level of at least d_{upper} is maintained, we feel 95 percent confident that 80 percent (P_0) of the lake users will be satisfied. When the covariate temperature (u) is also used in the model, we have provided values of d_{upper} for three selected values of u.

We conclude this section by making the following observations. While knowledge of d_{upper} may give us some idea of the level of public satisfaction, it is evident that natural and environmental factors may be present that preclude desired levels of d_{upper} from being maintained always. Nevertheless, this measurement and the correlation between the upper confidence limit and the degree of public satisfaction is important. It provides the Hillsdale agency with a guide to fulfilling consumer demand. In particular, should the upper confidence limit consistently fall below desired levels, the agency will know that recreational users are not satisfied, and presumably, will eventually cease using the facility.

Table 4.6

Summary of Hillsdale Lake Data

Day	Date	Ratio of 'good' responses of total	d	Weather	Temp.
1	6/21/99	84/105	0.93	Sunny	27.20
2	6/26/99	142/235	0.99	Sunny	32.22
3	6/29/99	88/126	0.95	Overcast	23.89
4	7/03/99	91/120	0.60	Sunny	31.67
5	7/07/99	41/41	0.91	Sunny	31.11
6	7/12/99	37/37	0.89	Sunny	26.11
7	7/15/99	29/36	0.93	Sunny	32.22
8	7/18/99	188/210	1.02	Sunny	33.33
9	7/23/99	34/35	1.03	Sunny	36.67
10	7/25/99	30/32	1.22	Sunny	37.22
11	7/26/99	57/60	1.52	Sunny	37.78
12	7/31/99	27/34	1.04	Rain	31.67
13	8/04/99	20/20	1.45	Sunny	30.00
14	8/08/99	97/111	1.17	Sunny	30.56
15	8/10/99	30/36	1.4732	Sunny	33.90
16	8/20/99	39/43	1.4605	Sunny	28.33
17	8/26/99	5/5	1.4478	Rain	24.44
18	8/29/99	50/64	1.6002	Sunny	31.67
19	8/30/99	11/11	1.5367	Overcast	28.89
20	9/03/99	37/39	1.1176	Sunny	32.22

Table 4.7

Analysis of Results by Logit Model without Covariate

	Logit model without covariate		
Parameter	All	Sunny	Overcast or rain
α	0.280	0.444	-2.962
$se(\alpha)$	0.341	0.322	1.693
β	0.940	0.824	4.025
$se(\beta)$	0.300	0.305	1.712
d_0	1.177	1.143	1.080
$se(d_0)$	0.091	0.104	0.060

Table 4.8

Ninety-five percent Upper Confidence Limits for Water Clarity

Temp. (°C)	All			Sunny			Cloudy or rainy		
	25	30	35	25	30	35	25	30	35
Logit w/o Temp.		1.327			1.315			1.178	
Logit with Temp.	1.797	1.431	1.230	2.459	1.612	1.340	1.217	1.204	1.313
Probit w/o Temp.		1.199			1.171			1.193	
Probit with Temp.	1.673	1.298	1.207	2.120	1.405	1.191	1.235	1.220	1.333

REFERENCES

Agresti, A. (1990), *Categorical Data Analysis*, New York: John Wiley and Sons.

Berkson, J. (1955), 'Maximum Likelihood and Minimum χ^2 Estimates of the Logistic Function', *Journal of the American Statistical Association*, 50, pp. 130-62.

Besag, J.E. and P.J. Green (1993), 'Spatial Statistics and Bayesian Computation (with Discussion)', *Journal of the Royal Statistical Society, Series B*, 55, pp. 25-37.

Cohen, A. and H.B. Sackrowitz (1984), 'Testing Hypotheses about the Common Mean of Normal Distributions', *Journal of Statistical Planning and Inference*, 9, pp. 207-27.

Fairweather, W.R. (1972), 'A Method of Obtaining an Exact Confidence Interval for the Common Mean of Several Normal Populations', *Applied Statistics*, 21, pp. 229-33.

Fisher, R.A. (1932), *Statistical Methods for Research Workers*, London: Oliver & Boyd.

Geman, S. and D. Geman (1984), 'Stochastic Relaxation, Gibbs Distributions and the Bayesian Restoration of Images', *IEEE Transactions, Pattern Analysis and Machine Intelligence*, 6, pp. 721-41.

George, E.O. (1977), 'Combining Independent One-sided and Two-sided Statistical Tests—Some Theory and Applications', Doctoral dissertation, University of Rochester.

Gilbert, R.O. and J.C. Simpson (1990), 'An Approach for Testing Attainment of Soil Background Standards at Superfund Sites', (Private Communication).

Hastings, W.K. (1970), 'Monte Carlo Sampling Methods Using Markov Chains and their Applications', *Biometrika*, 57, pp. 97-109.

Jordan, S.M. and K. Krishnamoorthy, (1996), 'Exact Confidence Intervals for the Common Mean of Several Normal Populations', *Biometrics*, 52, pp. 77-86.

Li, X. (2000), 'A Bayesian Analysis of Gasoline Data', Technical Report, Department of Mathematics / Statistics, University of Maryland Baltimore County.

Li, X., B.K. Sinha and B. Nussbaum (1999), 'A Statistical Analysis of Hillsdale Lake Data', (submitted).

Marshall, A.W. and I. Olkin (1967), 'A Multivariate Exponential Distribution', *Journal of the American Statistical Association*, 62, pp. 30-44.

Nussbaum, B. and B.K. Sinha (1995), 'Cost Effective Gasoline Sampling Using Ranked Set Sampling', Technical Report, Department of Mathematics/Statistics, University of Maryland Baltimore County.

O'Brien, R., B.K. Sinha and W.P. Smith (1991), 'A Statistical Procedure to Evaluate clean-up Standards', *Journal of Chemometrics*, 5, pp. 249-61.

Odom, B. and B.K. Sinha (1995), 'Asymptotic Semiparametric Procedures to Evaluate clean-up Standards', *Journal of Applied Statistical Science*, 2, pp. 89-102.

Patil, G.P. and C. Taillie (1990), 'Evaluating the Attainment of Interim Clean-up Standard', Technical Report, Center for Statistical Ecology and Environmental Statistics, Department of Statistics, Penn State University.

Sarkar, S.K. (1987), 'A Continuous Bivariate Exponential Distribution', *Journal of the American Statistical Association*, 82, pp. 667-75.

Shen, W.H. (2000), 'Exact and Approximate Tests in Multiple Logistic Regression Models', *Journal of Applied Statistical Sciences*, 9, pp. 183-92.

Sinha, B.K. (1988), 'Berkson's Bioassay Problem: Issues and Controversies', *Methods of Operations Research: Symposium on Operations Research*, 60, pp. 91-102.

Sinha, B.K. and B.K. Sinha (1995), 'An Application of Bivariate Exponential Models and Related Inference', *Journal of Statistical Planning and Inference*, 44, pp. 181-91.

Srivastava, M.S. and C.G. Khatri (1979), *An Introduction to Multivariate Analysis*, New York: North Holland.

Stouffer, S.A., E.A. Suchman, L.C. DeVinney, S.A. Star, R.M. Williams, Jr. (1949), *The American Soldier, Volume 1. Adjustment During Army Life.* Princeton, N.J: Princeton University Press.

Tippett, L.H. (1931), *The Method of Statistics*, London: Williams and Norgate.

Yu, P.L.H., Y. Sun and B.K. Sinha (1999), 'Inference About the Common Mean of a Bivariate Normal Population with an Environmental Application', Technical Report, Department of Mathematics/Statistics, University of Maryland Baltimore County.

5

Air Pollution

Statistics and Environmental Health Perspectives

PRANAB K. SEN

This interdisciplinary assessment of risk for environmentally induced adverse health conditions (diseases, disorders and mortality) aims to identify the contaminants and their impact by an understanding of the toxicity intake and the aftermath processes; statistical reasoning is indispensible in this context. Along with some recent developments, the merits and demerits of conventional statistical methodology are appraised.

1. INTRODUCTION

Environmental pollution (EP) and environmental toxicity (ET), taken together as EPT, is in a life-threatening phase; toxic and microbial, air, water, subsoil contaminations are escalating at an alarming rate, making our environment unsafe and unhealthy and tampering with our *quality of life* (Abdelhardt *et al.* 1985, Blot *et al* 1980 and Sacks and Steinberg 1994). Quantitative EPT risk assessment requires an interdisciplinary approach. *Gnotoxic, reproductive toxicity* (RT) and mental health problems have also been linked to EPT. The (carcinogenic) *greenhouse effect* is an overwhelming factor in the recent global environmental climatic and ecological imbalance.

In quantitative EPT risk assessment, environmental science, epidemiology and other public health disciplines which are a part of

---

I am grateful to J.K. Ghosh for his helpful suggestions and to R.F. Woolson for a good discussion of the IPGW dataset and his consent to my including a brief discussion here.

environmental health science (EHS) play a basic role; yet (bio-) statistics occupies a focal point. Statistical perspectives in EPT risk assessment relate to the following aspects:

(1) inventory of sources and quantification of EPT levels,

(2) dosimetric (and clinical) studies for biological effects, and

(3) observational studies to correlate the response and EPT matrices.

There is a natural emphasis on (1) from environmental science perspectives, and on (3) from EHS perspectives. The existing statistical literature relates mostly to the identification of EPT sources and their level of impact while epidemiologic (observational) studies relate to some modifications. The recent texts and monographs on environmental statistics (Barnett and Turkman 1997, 1994, 1993; Cothern and Ross 1994, Hallenbeck 1993, Hewitt 1992, Ott 1995, Patil and Rao 1994, Pearson and Turton 1993, Walden and Guttorp 1992) have a predominant flavour of (1) and (3), though there is a need to integrate (2) with (1) and (3). This has caught the attention of quantitative EPT analysts (Piegorsch and Bailer 1997 and Piegorsch and Cox 1996). For case studies in environmental statistics we refer to Cox *et al.* (1998). There is a need to incorporate more dosimetry in statistical analysis. The emphasis on conventional time series and spatial models (Smith 1989, Smith *et al.* 1998) and linking outdoor airborne particulate matter (PM) to daily mortality or other risk measures (Pope *et al.* 1992, Samet *et al.* 1997, 1995), Schwartz 1993, Schwartz and Dockery 1992, Schwartz and Marcus 1990, Styer *et al.* 1995), without accounting for the toxicity undercurrents, may not provide a complete picture. The distribution of EPT in the concerned population, their uptake distribution and demographic patterns have bearings on observational and epidemiologic studies.

Sources of environmental pollution and toxicity, human uptake and current statistical approaches to EPT risk assessment are oulined in Sections 2 and 3, and modelling and analysis is outlined in Section 4. In Section 5, along with some general remarks, a statistical appraisal of an epidemiologic study of Persian Gulf War (PGW) veterans is made.

2. EPT: Prevalence, Prognosis and Human Uptake

The current deleterious impact of EPT is mostly man-made. Airborne *respirable suspended particulate matters* (RSPM), toxic gases and fumes and other environmental toxins, often carcinogenic, have escalated health hazards. The *absorption* of environmental toxins is mostly through the skin with access to blood vessels. *Ingestion* relates mainly to the GI tract with access to the liver, kidney, mammary organ (in females), brain and developing organs (in fetus) through blood. *Inhalation* is the principal EPT source and takes place through the respiratory tract with access to blood and the organs mentioned before; virus progression is linked to blood and sexual contacts (Sen 2001 a, b, c; Sen and Margolin 1995). The EPT impact on human health and the ecosystem should be assessed while also considering xenobiotic effects.

Conventional measures like the *quality of air* or *ozone concentration* convey very little information regarding the xenobiotic effects. Most of the environmental regulatory agencies, such as the US Environmental Protection Agency (EPA), are coping with the immensurable task of assessing the sheer magnitude of environmental pollution while other health agencies such as the National Institutes of Health (NIH) and the National Institute of Environmental Health Sciences (NIEHS) in USA are assessing xenobiotic effects mostly through dosimetric (animal) studies and clinical trials. On the other hand, epidemiologic studies based on observational data require additional care when statistical reasoning is applied. Environmental and clinical epidemiology aim to deal with the composite EPT effects in a comprehensive manner. Dealing with the impact of specific EPT source(s), model identifiability issues may surface due to confounded factors (Breslow and Day 1987, 1980; Sen and Margolin 1995).

2.1 Greenhouse Effect and Ozone Threat

Due mostly to industrialization, the earth's ozone layer is thinning, resulting in the elevation of the topospheric ozone level (ultra-violet radiation), unprecedented climatic changes and carcinogenic effects (skin cancer) on exposure to direct sunlight or outdoor activity for a prolonged time.

2.2 Smog and Quality of Air

The air, heavy with RSPMs, toxic gases from industrial and automobile exhausts and waste products from a hazy smog layer, in conjunction with atmospheric humidity. It is a threat to human health. From epidemiologic studies, it is clear that the incidence rates of respiratory, skin as well as cardiovascular diseases have jumped to new heights. The *quality of air* index monitors the safety level for exposure to the outdoor environment; we need to pay more attention to its statistical assessment.

2.3 Environmental Smoking

Environmental smoking relates to active and passive smokers who inhale carbon mono-oxide (CO) and other toxic gases, and are susceptible to lung diseases (including tuberculosis and cancer). The cases of other narcotic drugs are quite similar, though instead of nicotine there may be the infusion of some other (and often more virulent) toxicants. Emission of smoke (containing carbon/sulphur particles and *CO*) has a similar toxicity (Blot *et al.* 1980, Doll *et al.* 1965, Doll and Peto 1978, Haenszel *et al.* 1962, Hirayama 1975 and Welsh *et al.* 1982).

2.4 Mineral Fibres and Coal Miners' Disease

Sub-soil mining and milling and drilling practices carry significant air pollution problems due to the release of various RSPMs; their inhalation results in lung blockage, leading to various lung diseases. As water is also contaminated, the pollutants may also be ingested or absorbed. The scenario is similar in the case of ***asbestosis***. Such mineral fibres, abounding in manufacturing sites and in places where such insulation materials are used, can cause lung diseases (Abdelhardt *et al.* 1985, Acheson and Gardiner 1980, Hobbs *et al.* 1980, Kaldor *et al.* 1986, Peto *et al.* 1982, Selikoff *et al.* 1980, and Welsh *et al.* 1982). Another notable example is the 'indigo workers' tale (Sen and Margolin 1995). In such cases, all the three uptake avenues can be identified (Berry and Wagner 1969). As for workers at dye-stuff manufacturing plants, tumor in the urinary bladder is prevalent (Case *et al.* 1954, Infantine *et al.* 1977, Lee-Feldstein 1983, and Whittemore and McMillan 1981).

2.5 Waste Products Disposal

Toxic waste from oil refineries, thermal power plants, nuclear power plants, and disposals from iron and steel plants, fertilizer, pest control products and other chemical plants, and even the garbage disposal problems are related to waste products disposal. Though the inhalation and absorption modes are overwhelming, ingestion mode is also pertinent.

It is clear that the mode of EPT intake by human beings is usually complex; and without toxicologic understanding, statistical conclusions based on observational studies may not be proper. Current advances in molecular biology and genetics have improved the understanding of EPT carcinogenicity. In a healthy human, somatic cells replicate and divide in an equilibrium; carcinogenic cells divide at an uncontrolled rate forming a tumor which may spread anywhere in the body.

Carcinogens including many RSPMs damage body cells and can eventually cause at least one cell to become cancerous, which over generations of cell divisions forms a 'mafia'. Interacting toxicants can accelerate the cancer formation process. Therefore, statistical appraisal should take into account all the three EPT factors.

3. Statistical Perspectives

The network of *E-maps* for measuring air pollution in different urban and other susceptible regions, designed to assess EPT intensity and diversity, is invoking increasingly sophisticated environmental statistical methodology. Since a complete EPT census is impractical, representative sampling schemes are advocated. *Length-biased sampling* (Patil and Rao 1978) and *ranked set sampling* (McIntyre 1952) schemes have been proposed in EPT studies (Muttlak and McDonald 1990, Patil *et al.* 1992). Typically, time-series data are collected at the chosen sites. *Quality of air* and *ozone concentration level*, common environmental safety regulatory indicator functions are applied. They relate to the level of RSPMs but may not reflect totally the impact on human health unless the biomedical factors are integrated properly. The ozone concentration level serves as a good warning with regard to the radiation effect of outdoor exposure to sun, due to thinning ozone layers. Whichever sampling scheme is selected, there is a need for spatio-temporal statistical modelling and analysis. Environmental statistical methods have been adapted mainly and analysis. Environmental statistical methods have been adapted mainly from *spatial processes, extreme value theory,* and *Markov fields*; they need to address spatio-temporal inhomogeneity and non-stationarity.

Designing the sites in the proximity of an EPT source is an important factor (Diggle 1990, Diggle *et al.* 1997, Stone 1988). Nevertheless, spatial inhomogeneity usually creates complications in statistical modelling (Guttorp and Sampson 1994, Le and Zidek 1992, Sampson and Guttorp 1992) and Markov fields have been suggested (Besag 1989, 1974; Besag *et al.* 1991); they involve spatial contours for formulating transition probabilities (incorporating, as needed, Bayesian methodology). In view of the usual high-level non-stationarity and possible inhomogeneity of environmental spatio-temporal fields, such formulations must be judged carefully. Semi-parametric models accounting for such non-stationarity, albeit retaining the Markov structure, are more flexible (Dey *et al.* 1998). The situation is usually different from standard linear models in spatial statistics (Cressie 1993). *Generalized Linear Models* (GLM), *Generalized Additive Models* (GAM) and non-parametric models may be more appealing.

The above developments relate mostly to EPT monitoring, their spatio-temporal distributions and the study of allied dependence patterns. EHS researchers, more concerned with the health impact of EPT and disease-disorder etiology, advocate dosimetric and toxicologic research to study the statistically possible adverse effects of EPT. *Environmental biometry* (Piegorsch 1994) has emerged as a viable discipline for the study and application of statistical methods in EPT studies on biological systems. Environmental, clinical and genetic epidemiology, devoted to the aspects of both toxicology and ecology, through observational (demographic and epidemiologic) studies, care for novel statistical methodology (Sen and Margolin 1995).

4. Statistical Assessment of EPT Risk

An assessment of EPT risk should take into account

(1) The statistical assessment of EPT and the uptake process,

(2) An understanding of the reactions of such EPT on the human body, mostly through bio-monitoring, and

(3) A quantitative formulation and interpretation of risk that explains the observed health effects.

Environmental pollutant concentrations and fluxes are usually measured at a number of grid-points (sites), though they are aimed as total loadings over areas defined by ecological or geo-political boundaries. The composite distribution of the EPT is determined by the spatio-temporal distribution of sources, along with relevant auxiliary variables. Information on uptake sources and their intake process distributions are valuable in bio-monitoring process that relate to factor 2. Response variables that come up in the form of identifiable health problems (diseases and disorders) are then used in factor 3 to assess the impact of EPT. It is difficult to arrive at factor 3 directly from factor 1, ignoring factor 2. Recently, in clinical epidemiology (involving human subjects), factor 2 is incorporated in a more effective manner to bridge this gap between factors 1 and 3.

In a conventional demographic set up, cause-specific mortality and morbidity rates are used to quantify the risk; in epidemiological set ups, such rates are modified further in the light of various other relevant explanatory (or auxiliary) variables. In bio-medical studies, the risk is often interpreted in terms of the survival function $S(t)$ or equivalently the hazard function

$$\lambda(t) = -(d/dt)\log S(t)$$

which is also known as the ***instantaneous mortality rate***. In the presence of (a number of) explanatory variables, the conditional survival or hazard functions, given the explanatory variables, is used. We denote the conditional hazard function, given

$\mathbf{Z} = \mathbf{z}$ by $\lambda(t \mid \mathbf{z})$

which, in a parametric framework, can be expressed as a known function of some unknown parameters. Thus, the risk can be interpreted in terms of these parameters. There may be limitations due to considerations of validity and robustness, particularly because there are many explanatory variables, not all being continuous or even quantitative. Semi-parametric and non-parametric methods have therefore been proposed to induce more robustness in a semi-parametric model. Cox (1972) proposed the model:

$$\lambda(t \mid \mathbf{z}) = \lambda_0(\mathbf{t})\exp\{\beta'\mathbf{z}\} \qquad (5.1)$$

where $\lambda_0(t)$ is non-parametric, while the hazard regression is parametric involving a finite number of parameters. This way, one eliminates the lack of robustness in one direction, though the parametric component may still induce non-robustness. Various complications arise when there are multiple end-points and/or the covariates are subject to measurement errors or misspecifications. Some of these problems in a biometric context are discussed in Clegg *et al.* (1999); other pertinent references are cited there. In the EPT context, we need to appraise this situation with due attention to data quality and extractable statistical information. In large-scale experiments, a non-parametric model can be chosen to achieve robustness in a global sense, though its efficacy for finite sample sizes may be low.

Such conditional hazard or survival functions rest on the basic assumption that the response as well as auxiliary variates are defined and observable for each individual unit. This ideal situation is rarely tenable in EPT risk analysis; the auxiliary or factor variables, such as the EPT, may not be precisely measurable (or even identifiable). Also, their uptake by the concerned population is not uniform. The pattern of exposure levels and other relevant factors have a stochastic impact on the dose-response regression. More crucially, the response and auxiliary variables, being typically population-based, may not relate to individual subjects. These deviations make it dificult to adopt the survival function approach to EPT risk analysis. The following two lines of attack constitute the main development:

(1) Detailed spatio-temporal modelling of EPT sources and their prevalence, and

(2) Refined statistical modelling of observational studies with pertinent EPT factors in a GLM/GAM.

According to this perspective, validity and efficiency aspects dominate statistical prospects while the incorporation of toxicologic findings can be statistically rewarding. Given the volume of accomplished statistical research in the areas 1 and 2 (the bibliography in Piegorsch *et al.* 1997) and their adherence to areas 1 and 2, we shall only make a few pertinent comments on their state of the art, and confine ourselves to some other recent developments. We may conceive of a typical spatio-temporal process

$$\mathbf{X} = \{X(s,t), s\varepsilon S, t\varepsilon \tau\}$$

where both the state and time domains are in continuum, and where

$\mathbf{X}(s,t)$ is itself a vector of usually high dimension. In the process of actual data collection, a discretized version of the space and time domains is generally adopted, so that we have a network of points (location)

$s_j, j = 1, \ldots, K$ in the space domain S, as well as time-points

$t_j, j = 1, \ldots, M$ in the time domain τ, and our observable data set relates to

$$X(s_j, t_i), j = 1, \ldots, K, i = 1, \ldots, M$$

where it is tacitly assumed that these points are dense in $Sx\tau$. Here, the ordering of the time-points t_i creates no conceptual difficulty while in general it may be difficult to induce an (even partial) ordering on S. For a given t_i, the collection

$$\{X(s_j, t_i), j = 1, \ldots, K\}$$

represents a point process, though in EPT studies the s_i may not be dense on S. Moreover, the spatio-temporal dependence pattern might not be known or guessed well. There have been some recent texts and monographs that contain some up-to-date developments of Gaussian as well as non-Gaussian processes with spatial structures; we refer to Rosenblatt (1999) and Stein (1999). However, what we need is a *kriging* model by which we can estimate validly and efficiently $X(s,t)$ at non-grid points as well. Essentially, in a kriging approach, one takes a linear combination of neighbouring grid-point observations with due consideration to their dependence pattern. In kriging methodology, the formulation of a *variogram* is an important task. In a stationary case, we can define a variogram by the covariance

$$\begin{aligned}\Gamma(u,v) &= E\{[X(s+u,t+v) - X(z,t)]\times \\ &\quad [X(s+u,t+v) - X(s,t)]'\},\end{aligned} \tag{5.2}$$

which remains the same for all s, t. However, as the EPT spatial processes are mostly non-stationary (Carrat and Valleron 1992), we may need to consider an extended definition where $\Gamma(u,v)$ depends on s, t as well. This might take away the charm and simplicity of the kriging approach. In this respect, linearity as well as weakly second order spatial stationarity are presumed (Mathéron 1973). This assumption is likely to be untrue for higher dimensional data clouds, especially in EPT contexts. Linear interpolation in the kriging methods also needs an appraisal for non-linear spatial processes, especially when the $X(s,t)$s are vector values and not all the components have continuous variables. In spatial interpolation, polynomial regression analysis, under the name of *trend surface analysis,* has also been advocated; this also has affinity to linear models that may not be appropriate in EPT studies. For these reasons, conventional linear models have been replaced by GLM/GAMs that are more applicable for discrete variables. In particular, Poisson regression models have been advocated by a host of researchers (Datta *et al.* 2000, and others). It may be remarked further that even for GLM/GAMs, usually we do have a (multivariate) *mixed model.* In the process, the basic simplicity of a GLM approach is compromised and we may need to appeal to *quasi-likelihood* methods (Wedderburn 1976) and their counterpart, *generalized estimating equations* (GEE) (Zeger and Liang 1986). Even so, here the covariance matrix may no longer be simple enough, resulting in lesser appeal for GEEs. The case of extreme value theory in EPT studies seems to share some of these drawbacks. Justifying the relevance of extreme-value distributions for the error component and handling serial or cross-sectional dependence patterns may be difficult. The mixing-sequence and *long-memory dependence* patterns relax the assumption of independence to a certain extent; sans stationarity these models may become highly complex. While these modifications may suit temporal variation, in EPT studies we not only have a network of sites, but also at each site, there are multivariate observations with possibly discrete (or count) variables. Therefore, the extent of the adaptability of the extreme value theory in EPT spatio-temporal models would be enhanced as the basic theory is extended to cover more general dependent, discrete and multivariate cases.

At this stage, spatial GLM/GAMs that de-emphasize linearity seem to have a greater scope for adoption in environmental studies. Besag's (1974) innovation of the *auto-logistic* model in a simple spatial set up is a notable example in this respect. Consider a set of variables

$$X(s_j), j = 1, \ldots, K$$

where s_j refers to the grid-points in S. For any given s_j, consider the conditional distribution of $X(s_j)$, given all the other $K-1$ variables, $X(s_\kappa), \kappa \neq j$. If this conditional distribution depends functionally on a specific $X(s_r)$, then s_r is called a ***neighbour*** of s_j; also

$$K_j = \{r : s_r \text{ is a neighbour of } s_j\}$$

is termed a neighbourhood of s_j. Then, the conditional distribution of $X(s_j)$, given all the others, can be expressed in terms of the conditional distribution of $X(s_j)$, given only the neighbours:

$$\{X(s_r) : r \epsilon K_j\}$$

This leads to a simplified formulation of a pseudo-likelihood function that is amenable to statistical analysis. This formulation works out well in the presence of other covariates. Also, in the spatio-temporal scheme, it works out under some additional (mild) restrictions. This formulation is specially useful for binary outcomes. If we could view this as a multivariate time series model, though with possibly binary or categorical response variables, it may be tempting to appraise this as a *Markov chain* where the Markovian property is applicable not only to the temporal factor but also with respect to the neighbours when an ordering can be induced on them. In some spatio-temporal models, a neighbourhood may be defined in terms of the adjacent grid points (in two or three dimensions), so that this approach is adoptable. However, there are usually some limitations of pseudo-likelihood modellings (that include marginal, conditional, partial, and profile likelihoods) that are based typically on likelihoods for a subset of parameters. As is the case with partial likelihood, there may be a situation where we have a parameter, say θ of usually a small dimension, in conjuction with another parameter, say ν of usually a high dimension (it may even be a functional parameter). Our main interest is to formulate suitable likelihood-based inferences about θ when ν is unknown. All these variants of pseudo-likelihood modellings are based on suitable transformations, partitioning of the observed stochastic elements and eliminating a part so as to be able to use simpler likelihood models for the part of interest. There are some conceptual impasses some of which have been eliminated only under some asymptotic setups. In many cases, these result in a slower rate of convergence and comparable rates for bias terms. For example, in a profile likelihood approach, the choice of the penalty function requires delicate attention with special reference to the contemplated applications, and in EPT studies, that may indeed be a delicate task. Ghosh and Rao (1994), mingling (empirical and hierarchical) Bayes with frequentist method-

The approach considers the neighbouring grid-points, puts plausible priors on their dependence pattern, and as in a variogram approach, utilizes this additional statistical information in an empirical or hierarchical Bayes set up to borrow strength from neighbours in order to estimate at a grid-point in an improved way. In the context of the analysis of geographical variation of disease risk, Bernardinelli *et al.* (1997, 1995a, b, 1992) and others have considered empirical as well as fully Bayesian methods. Though in this context, ET information has not been utilized fully, the employed empirical and hierarchical Bayes methodology seems to have good prospects, especially when in the choice of priors, ET background information is incorporated properly.

We may remark that because of the extremely high dimensionality as well as spatio-temporal nonstationarity, there are some genuine statistical concerns for incorporating point processes in EPT risk assessment. Semi-parametric point process models have also been considered in the recent literature. Though such models allow more flexibility (relative to parametric models), there remains the (often formidable) task of ascertaining the validity of the non-parametric component in such semi-parametric models. Secondly, since both the state and time domains are continuous, a discretized version of the space and time domains is essential for data collection. Thus, we generally have a very high-dimensional point process, with an unknown, complex dependence pattern. This feature, coupled with the possibly non-Gaussian nature of these high-dimensional processes, may make it difficult to adopt standard statistical methodology for such EPT studies. As a viable alternative, Bayesian spatial methods have evolved in this environmental epidemiological set up (Datta *et al.* 2000, Ghosh *et al.* 1999, Lawson and Cressie 2000, Marshall 1991, Waller *et al.* 1997 and Zia *et al.*). There is, however, a need to extend the existing methodology to cope with non-homogeneous spatio-temporal dependence patterns that are encountered commonly in practice. In a stationary spatial process, the variogram serves a basic role. For non-stationary spatial processes, the estimation of a variogram from observed data may be more difficult. The Bayesians have tried to resolve this issue in a somewhat different way (Le and Zidek 1992). But, the usual type of priors that are incorporated in such empirical and hierarchical Bayes models may need to be appraised for explaining well the underlying spatial and temporal dependence patterns. In this respect, they share a similar drawback with the Kriging methodology. *Markov Chain Monte Carlo* appraised for explaining well the underlying spatial and temporal dependence patterns. In this respect, they share a similar drawback with the Kriging methodology. *Markov Chain Monte Carlo* (MCMC) (iterative) methods (Gelfand and Smith 1990, Gelman *et al.* 1995) have been used more aggresivly in the Bayesian context to simulate stochastic processes; and these methods are useful in EP studies as well. The *Gibbs sampling* scheme (Geman and Geman 1984), and the Metropolis–Hastings algorithm (Hastings 1970) can be incorporated to carry out the simulation work in a more effective way, incorporating Bayesian concepts and amending them in the light of environmental constraints. In current EPT studies, there is ample room for further developments. Non-parametric and semi-parametric approaches may generally fare better, though they may generally need larger data sets. We may also remark that because of possible genetic undercurrents, environmental and genetic epidemiology need to be appraised in a common vein, and there are still some statistical challenges.

We now focus on dosimetry and clinical epidemiology in EPT studies. As stated earlier, for EPT analysis, though our target population relates to human beings, dosimetric studies involve (and often involve only) sub-human primates, possibly at quite different dose and exposure ranges. As dosimetric studies allow greater experimental control, involving simpler experimental designs, the rationality and validity of extrapolation from sub-human primates to humans need to be appraised properly. Mainly, dosimetric studies provide scope for the incorporation of a bio-toxicological background to a greater extent, enhancing the relevant statistical information. Consider the environmental smoking problem. For a target population, one needs to assess firstly the prevailing level of EPT and secondly, the distribution of the exposure level of concerned humans to such EPTs. Since the EPT intake dose is highly dispersed, a formulation of statistical regression models needs to take into account its impact on the process of biological reactions and health hazards; conventional dose-response regression models, which ignore these features, are not appropriate here. In a dosimetric study, the emphasis is on modelling the bio-toxicological process, so that if a comparable model can be conceived for human intake, then this additional information could be utilized in improved risk analysis. The PBPK modelling in toxicologic studies is based on the latter motivation. Based on physiological backgrounds, a pharmacokinetic model is often used to chart the flow of events leading to the response variables. Once the pharmacokinetics can be depicted in a mathematical (though stochasic) model, the wealth of statistical tools can be used to draw statistical conclusions in a more assessible manner. The most noticeable negative aspect of the PBPK modelling is the fact that the pharmacokinetics in sub-human primates may not be close enough to that in humans so as to extrapolate conclusions. Recent developments in pharmacokinetics have been instrumental in bridging this gap, though much more remains to be accomplished. The health impact of EPT is undergoing evolutionary appraisal from a broader perspective. Reproductive toxicology (RT) has captured the attention of epidemiologists who are correlating its salient features to environmental toxicology as well as to *Occupational Toxicology* (OT), though their studies are mostly observational in nature. The EPA has set some guidelines for risk assessment of RT that includes any adverse effect on the male, female or couple resulting from exposures which produce alterations in sexual behaviour, pregnancy fertility, pregnancy outcome or other functional changes dependent on the reproductive system. Male as well as female fecundity, as defined and interpreted in a biologically meaningful way, lies at the center of RT risk assessment. In order to assess the impact of environmental exposure, bio-markers are usually used as surrogate endpoints (Mazumdar *et al.* 2000). The incorporation of ET in this set up has already started with dosimetric studies involving rodents and sub-human primates. The statistical findings from such animal studies should provide valuable guidelines for clinical trials involving human subjects. These findings can then be used for pooling information from ET studies with epimediologic studies. Otherwise, there remains concern for confounding conclusions derived solely from observational studies and their lack of validity.

Some of the grave effects that EPT has on the quality of life are being appraised. Chronic diseases, partiularly respiratory ones, are surfacing with much higher incidence rates in geo-political and ecological regions having identified EPT concentrations. Cardiovascular mortality rates have been linked to EPT. Depression and related psychiatric problems also have affinity to EPT. The aging process in human beings is also dependent on EPT. Current research in these areas aims to focus on the biological impact of EPT at the molecular level; this added information needs to be incorporated in statistical modelling and analysis. In that context, the PBPK and bio-mechanistic modellings are very important.

The impact of EPT on human population is apparent from the

elevated risk for cancer (of various types), chronic diseases and disabilities that lead to the deterioration of the quality of life in a broader sense. Led by early research on statistical modelling of human carcinogenesis (Armitage and Doll 1957, 1954; Neyman and Scott 1967, Stocks 1953, and others) with due emphasis on appropriate bio-mechanistic models (though without a proper understanding of cancer etiology), recent developments in cancer biology have broadened our understanding of the progression from normal cellular functioning to malignancy. It is conceivable that there could be a transition from normal cells to initiated ones; and then from the initiated ones to malignant ones, where of course, in each phase a cell can divide into two cells, or die out. In the simple case (i.e. without malignancy), suitable Markov (e.g. branching) processes can be adopted to formulate the transition from one cell to two daughter ones and its alternate, death. These transitions take place subject to the natural equilibrium and they are repeated over generations of cell divisions. In some simple cases, if stationarity prevails, the probability of extinction (over generations of cell divisions) can be obtained explicitly. Such models have been extended to accommodate malignant cell formation along with the absorbing stage, death. The transition from a normal cell to an initiated one is to be depicted with a possibly different transition probability; and there also, transitions to division into two cells or death are to be formulated in a Markovian set up. Finally, from an initiated cell, the transition to a malignant one can be depicted in a Markovian model with a possibly different transition probability and allowing a (stochastic) number of cell divisions at the initiated stage. For both the transitions of normal to initiated cell and from initiated to malignant cells, the PBPK modelling plays a basic role. The ET impacts can then be incorporated in such models by assigning higher transition probabilities to the transition from a normal cell to an intiated one as well as from an initiated one to a malignant cell. Therefore, one might consider a three-stage stochastic model akin to the classical Neyman-Scott one, but incorporating the toxicants as covariates, with plausible regression structures. This model can be supplemented with PBPK interpretations for the growth and colonization of malignant cells. This development led in turn to the multi-stage theory, as has been advocated by a number of researchers. Whittemore and Keller (1978) reviewed quantitative theories of carcinogenesis. Kohn *et al.* (1993), Moolgavkar *et al.* (1992, 1981) and Portier and Kopp-Schneider (1991) have considered more general multi-stage models;

Portier and Sherman (1994) used multi-stage models to study the potential effects of chemical mixtures on the carcinogenic process. Mathematical modelling of carcinogenesis in primates involves modelling of several aspects:

(1) exposure,

(2) dose delivery/tissue dosimetry, and

(3) cellular kinetics.

The latter concerns the pharmacodynamics by which cellular alterations result in uncontrolled cell growth which in turn leads to the formation of tumors. Mutation and cancer can be linked in this way. However, not only is the cancer etiology not precisely known, but there might also be a vast number of auxiliary or explanatory variables that have some bearing on the anticipated carcinogenic process. To account for this overwhelming stochastic cloud over such bio-mathematical models for carcinogenesis, stochastic modelling includes due stochastic components in addition to the deterministic ones, amenable to suitable goodness-of-fit tests. Considerable research work in this direction has been carried out at the NIEHS (in collaboration with other research centres). We refer to Kopp-Schneider *et al.* (1998), and El-Masri and Portier (1999) where other pertinent references are cited. The latter paper deals with a mathematical model, and the former one with possibly non-homogeneous Poisson processes in a multi-stage set up. Nevertheless, such bio-mechanistic models have mostly been considered for sub-human primates and living cells (or tissues) where PBPK explanations generally work out well. It remains to be seen how such PBPK-enriched bio-mechanistic models can be implemented effectively for ET studies involving human beings.

Toxicokinetics and biochemical modelling is a fundamental task in the assessment of EPT risks. Such models attempt to incorporate mathematical and statistical tools in conducting biological studies to facilitate the formulation of valid dose-response models. Toxicokinetic data relate to absorption, distribution, metabolism and elimination of environmental agents while mechanistic data relate to tissue response, for example, mutagenicity and altered gene expression. There is a genuine need for the development of methods to characterize the relationship between exposures to hazardous environmental agents and the induction of adverse biochemical effects. Dosimetric (animal and tissue) studies are employed commonly to probe into this aspect, and more and more complex studies are being undertaken to further our understanding of the basic principles of toxicokinetics and pharmacodynamics in the case of EPT progressions. In EPT risk analysis, cause-specific mortality and morbidity rates dominate the picture whereas in toxicokinetics, because of the complexities of the response variables, a different kind of statistical formulation seems to be necessary. In conventional dosimetric studies, the basic principles of biological assays govern the statistical methodology, though in EPT risk analysis, as has been mentioned earlier, there are some impasses that need to be examined carefully. These are dicussed below.

4.1. Precise Assessment of 'Dose' in EPT Studies

It is generally assumed that although the response variables are stochastic, the auxiliary variables can be measured without error. In EPT studies, to the contrary, even in laboratory dosimetric studies, there are *measurement errors* associated with administered doses. They are naturally different from conventional linear models with measurement errors (Fuller 1986). In epidemiologic studies, the disease-exposure relationship, if subject to measurement errors on the exposure variables, can be confounded; and the extent of damage depends on the level of measurement errors. The situation is worse for binary or count-response variables. With that in mind, a variety of measurement error models in dosimetric studies have been considered by a host of researchers. If $\mathbf{Y}$ stands for the response vector while $\mathbf{x}$ for the dose or exposure variables, it is customary to formulate a regression:

$$\mathbf{Y} = \psi(x, \theta) + \epsilon, \tag{5.3}$$

where ϵ stands for the error, and $\psi(\cdot)$ for the regression function. In a parametric formulation, the nature of $\psi(\cdot)$ is assumed to be known, apart from a number of unknown parameters, θ. Suppose now that instead of the true $\mathbf{x}$, we observe (or record) some other variables $\mathbf{W}$. In the simplest case, we may observe $\mathbf{W}$ such that $\mathbf{W} = \mathbf{x} + \mathbf{e}$ where the errors $\mathbf{e}$ are independent of ϵ and have null medians. If we substitute $\mathbf{W}$ for $\mathbf{x}$ in the regression function $\psi(\cdot)$, there may be bias and other distortions to the model. In that sense, $\mathbf{W}$ qualifies as a surrogate measure for $\mathbf{x}$, if the conditional distribution of $\mathbf{Y}$, given both $\mathbf{x}$ and $\mathbf{W}$, coincides with the conditional distribution, given $\mathbf{x}$ alone. This ideal case of *non-differentiable errors* may not generally dom variables with Poisson distributions $P(\lambda_i), i = 1, \ldots, n$, where the parameters λ_i stand for the means, then we may consider the canonical link functions

$$\theta_i = \log \lambda_i = x_i'\beta, i = 1, \ldots, n, \tag{5.4}$$

where the vectors x_i correspond to explanatory or auxiliary variables, and β refers to the regression parameter (vector). In this set up, we have a conventional log-linear model for the Poisson counts. However, in the present context, the vector x_i may not be observable, and the observable random variables $\mathbf{X}_i$ can be expressed as

$$\mathbf{X}_i = x_i + \epsilon_i$$

where the ϵ_i stand for the (unobservable) measurement errors. In such a case, using $\mathbf{X}$ instead of $\mathbf{x}$ in the link function results in a model distortion, even when treated in a conditional set up. The nondifferentiable clause may not be tenable here. A similar situation arises if we treat M_t as Bernoulli random variables with appropriate sample size N_i and probability $\pi(x_i)$, following a conventional logistic model. In that case we would have the logit model. This way, GLM or even GAM can be incorporated in EPT studies when the $\mathbf{x}_i$s are observable. However, such models need modifications when there are measurement errors. The evolution of beta-binomial or Dirichlet-multinomial distributions has patched up a part of this methodologic problem, though that may lead to an over-dispersion phenomenon. The use of more flexible (empirical/hierarchical) Bayes methodology is quite promising. In conventional linear and GLMs, under additional regularity assumptions, a statistical analysis of measurement error models has been developed (Fuller 1986, McCullagh and Nelder 1989). For some semi-parametric models (including the proportional hazards model, Cox 1972), similar measurement error models have been considered by a host of researchers (for example, Hougaard 1998, Murphy 1994). In order to adopt some of these models for actual EPT risk analysis, some difficulties may be encountered. Most notably, in many EPT studies, the EPT level measured from an EPT source as a whole might be quite different from the levels actually intaken by specific subjects or individuals, even if they belong to the same population. For example, in the ozone concentration problem, the atmospheric level determined by environmental scientists (at some observation post) may not truly reflect the actual level to which an individual is exposed. His/her exposure (duration, time, etc.) to the sun might be a major factor in the mensuration of the actual dose intake. So, when an epidemiologist wants to relate the incidence rate of skin cancer to the recorded ozone concentration level, it might be questionable whether or not the measured atmospheric level reflects the actual absorbed level by an individual, even if proper adjustments are made for duration/timing of exposure on a population basis. To put in a statistical framework, suppose that d_t stands for the ozone concentration level on day t at a particular site (say, a city) and that M_t stands for the daily mortality rate due to skin cancer in the same geographical area. There have been some epidemiologic efforts in correlating M_t with d_t with support from statistical reasoning (Schwarz 1993 and its follow-up studies). For example, for the population size N_t at time t, we could consider a Poisson regression model for

$$X_t = N_t M_t$$

where the parameter $\lambda_t = \lambda(d_t)$ depends on the population size N_t and the dose level d_t. We may then consider

$$P\{X_t = \kappa\} = e^{-\lambda_t} \lambda_t^k (k!)^{-1}, \ k \geq 0. \tag{5.5}$$

Often, we let

$$\log \lambda(d_t) = \alpha + \beta d_t, t \geq 0$$

so that β reflects the regression effect of the ozone concentration. In many cases, d_t is expressed as an ordered categorical variable (e.g., green, orange and red for safe, precautionary and alarming levels) and this may lead to variation in this modelling (measurement error might lead to misclassification of the state). This model treats all the subjects to be homogeneous with respect to the dose level. In formulating such a standard dose-response regression model in such a context, there are certain limitations. First, the pattern of exposure level to the outdoor UV-rays has a significant role in assessing the effective dose; even so, it could be influenced highly by occupational and other factors. A stratification of the population into relatively more homogeneous groups (with respect to the exposure level) would provide a better resolution. Secondly, skin cancer may not be an instantaneous response. It needs an incubation period and other biological factors whose impacts are observable over a time period. There could be a latent effect in the dose. As such, instead of other biological factors whose impacts are observable over a time period. There could be a latent effect in the dose. As such, instead of choosing d_t as an explanatory variable, it might be more reasonable to take a composite variable that reflects the cumulative effects over a period of time and incorporates other auxiliary variables that are more relevant biologically. Conditions such as sunstroke might be more related to the current level and for them a different model may be in order. A more serious problem is the lack of independence of the observable events for various values of t, particularly when they are not far from each other. In that way, instead of independent Poisson distributions for different t, we may need to bring in suitable counting processes that account for temporal dependence. In any case, as the values of d_t are not only stochastic but also subject to measurement errors, modelling becomes enormously complicated; there is a compliance factor described below that can describe the exposure dose intake process better. In passing, we may remark that Markov field models for the counting process encountered in temporal studies have a greater scope than simple GLMs, although non-stationarity of the transition probabilities need to be handled carefully.

4.2 *Stochastic Compliance Error Models*

We illustrate this with the compliance error logistic model. In a logistic model, at a dose level d, the binary response variable Y_d has the probability structure

$$\begin{aligned} \pi(d) &= P\{Y_d = 1|d\} \\ &= 1 - P\{Y_d = 0|d\} \\ &= \{1 + \epsilon^{-\alpha-\beta d}\}^{-1},\ d \geq 0 \end{aligned} \quad (5.6)$$

In a stochastic compliance model, the actual intake dose D is stochastic and we consider

$$D = U \times V, U \in [0,1], V \geq 0$$

where V stands for the administered dose level, and U for the proportion of that which results in the actual intake level D. We need to extend the classical logistic model in the light of the extra statistical variation that is due to U. In this respect, the distribution of U provides additional information that can be incorporated in the modelling process via a conditional logistic model. The extended logistic model obtained by integrating over D for an administered level V, though mathematically not so tractable, can be formulated. Note that

$$\begin{aligned}\pi^*(v) &= P\{Y=1 \mid v\} \\ &= E[P\{Y=1 \mid D, v\}] \\ &= \int_0^v \{1+e^{-\alpha-\beta d}\}^{-1} dF_{D|v}(d),\end{aligned} \tag{5.7}$$

where $F_{D|v}(d)$ stands for the conditional distribution of D, given V. Then, assuming *non-differentiable errors*, we have

$$P\{Y=1 \mid D, V\} = P\{Y=1 \mid D\}$$

and this leads to the final step from the intermediate one. In general, this expression may not have an algebraic form and much of the simplicity of the logit model is gone. Fortunately, in many EPT studies, statistical information on the distribution of U can be obtained from independent studies that can be used to provide suitable approximations to the corresponding unconditional model. Chen-Mok and Sen (1999) incorporated a general class of beta distributions (including unimodal, positively and negatively skewed ones) for U and suggested suitable modifications to statistical analysis schemes based on the logistic model. Note that if the compliance factor U has a beta distribution with parameters (a, b) (both non-negative), then

$$\begin{aligned}E(U) &= a/(a+b) \\ &= \gamma \text{ say}, Var(U) \\ &= ab/\{(a+b)^2(a+b+1)\} = \Omega \text{ say},\end{aligned} \tag{5.8}$$

where γ, Ω are non-negative functions of (a, b). The distribution of D (and its moments) for a given V can be readily obtained from the beta distribution of U. The advantage of using a beta distribution is its flexibility in modelling different patterns of compliance. Different combinations of (a, b) lead to a range of compliance patterns, that is, $a < 1 < b$ $(b < 1 < a)$, for low (high) compliance of dose, while for both a and $b > 1$, we have a bell-shaped distribution. Symmetric when $a = b$, it becomes more and more peaked as a and b increase.

In this context, it may be observed that in many EPT studies, there is a prevalence of very low dose levels, so that a Taylor series approximation (up to the second order) for $\pi^*(v)$ can be advocated with confidence. With this approximation, re-parametrization works out well. Then, the standard maximum likelihood method can be used with some modifications (Chen-Mok and Sen 1999). A very similar situation arises in radio-immunoassays where only a small radioactive dose is administered for bio-monitoring. In this general set up, re-sampling plans can also be incorporated in drawing statistical conclusions. Such compliance error models also amend to Bayesian analysis in a more adoptable manner than the measurement error models. Basically, in an empirical (heirachical) Bayes model, suitable priors are attached to the parameters α and β in the original logistic model. The posterior distributional parameters are then estimated from the observed data set using a combination of the frequentist as well as the Bayesian approach. Of course, in such a case, a judicial adoption of such priors is a prerequisite for the goodness-of-fit of such a Bayesian paradigm. We refer to Zeger and Karim (1991) where a Gibbs sampling approach has been proposed for mixed-effects GLM; see also the monograph, edited by Dey *et al.* (1998). There is, however, a pressing need to consider such modelling under a spatio-temporal set up, where for non-linear, GLM or non-parametric models, handling spatial as well as temporal dependence patterns could be more delicate. For example, working with the compliance error model, a natural set up would be to use a multivariate logistic model with stochastic dose factors. But such a formulation would depend on the extent to which we can incorporate the spatial dependence and temporal dependence factors in such a generally complicated multivariate model. The situation is even more cumbersome with non-parametric or semi-parametric modelling. In dealing with a single time-point event (as was the Hiroshima atom bomb episode), a spatial model is more appropriate, and we may refer to Chen-Mok and Sen (1999) for some (albeit partial) resolutions.

4.3. Extrapolation (from Mice to Man!)

Dosimetric studies are akin to bioassays, although there are problems here when possibly very low doses are used and with the extrapolation of the findings from sub-human primates to human beings. For example, dioxin-like compounds are ubiquitous environmental contaminants; their persistence in the environment, their lipophilicity and bio-accumulation (mostly through the food chain) may result in a chronic lifetime low-level human exposure. Although both sub-human primates and human cell systems respond to dioxin, their intake, metabolism and response mechanisms may not be isomorphic.

A causal relationship between dioxin exposure and health effects has been established in rodent models (see Walker *et al.* 1998 where other pertinent references are cited). In view of the basic differences in the bio-accumulation process, metabolism and exposure levels, there is considerable scientific controversy over the potential use of these findings for human health risk that is generally caused by persistent, daily, low-level exposure. To put in a statistical framework, suppose that for rodents, the prescribed (normal or high) dose level is D_0 and the (normal or shorter) exposure time is T_0, whereas for potential human risk, the (low-level) dose is D_1 and (much longer) exposure time is T_1. The effective dose for the rodent and human exposure to dioxin may then be taken (in a multiplicative set up) as

$$E_R = D_0 \times T_0, \text{ and } E_M = D_1 \times T_1. \tag{5.9}$$

The conclusions derived from a rodent study at effective dose E_R are intended to draw parallel conclusions for human exposure risk, with a simple adjustment for the difference in the two dose levels, E_R and E_M. The basic statistical question is: can we draw statistical conclusions from the rodent studies at level E_R as appropriate for human risk at level E_M? Not only may E_R and E_M vary considerably, but the multiplicative formula for the effective dose may also not be appropriate. In order to de-emphasize the multiplicative effect, one may consider a two-factor set up wherein the component (dose $\times$ response interaction) is included to account for a plausible deviation. Even so, typically, for sub-human primates we have moderate dose and moderate exposure, while for humans, the exposure is prolonged and the dose is generally very low. Therefore, the statistical issue is: can we draw statistical conclusions about the tail behaviour from data in the central part of a distribution? In a typical dose-response model set up, not only are the two design-points $D_j, T_j, j = 0, 1$ for the rodent and human exposure far apart, but, in view of the species difference, using a common form of the tolerance distribution (with possibly different parameters) may also be questionable. NIEHS and other environmental health agencies are probing into this branch of research to fill this knowledge gap by characterizing key events necessary for understanding the intricate mechanistic relationship as well as adverse biological responses. PBPK models have gained popularity in this respect. Biological factors are taken into account in the biomathematical modelling with provisions for stochastic analysis. At the present time, *stochastic partial differential equations* (SPDE), incorporating some of the pertinent biological factors, have been advocated mostly, wherein the Fourier coefficients are made to depend on biological understanding. However, there remains ample room for further development.

For a better understanding of such dosimetric studies, let us refer briefly to their precursors, namely, the biological assays designed to study the relative potency of a new drug with respect to a standard one; *parallel line* and *slope-ratio* models are used commonly. In *bio-equivalence models,* degined to study the equivalence of two or more different forms of the same drug, there are certain modifications, though the basic structure is retained. Bioassays are either of the direct type or the indirect type. In the context of EPT studies, we encounter indirect quantitative or quantal assays mostly. In such formulations, the basic tolerance distribution plays a fundamental role. In most bioassay models, logistic or normal distributions are adopted. In EPT studies, one of the basic problems is to identify the form of tolerance distributions so as to incorporate the logit or probit models for suitable statistical analysis. Transformation of dose and response variables can often make the tolerance distributions fairly symmetric; nevertheless, assuming them to be normal or logistic may not be very wise, particularly when the administered dose levels are not in the central part of the spectrum. Non-parametric and semi-parametric methods fare better in this context. In a parallel-line (indirect, quantitative) assay involving a standard and a test preparation (denoted by S and T, respectively) with tolerance distributions $F_S(y \mid x)$ and $F_T(y \mid x)$ respectively, where x stands for auxiliary or explanatory variables including the dosage level, it is assumed that

$$\begin{aligned} F_S(y \mid x) &= F(y - \beta'\mathbf{x}), \\ F_T(y \mid \mathbf{x}) &= F(y - ln\rho - \beta'\mathbf{x}), \end{aligned} \tag{5.10}$$

where F is a continuous distribution function, ρ stands for the relative potency of the test preparation with respect to the standard and β is the vector of regression parameters. In parametric models, F is usually taken to be a logistic or normal distribution. Finney (1964) has an excellent treatise of parametric bioassays. In a non-parametric and semi-parametric set up, through Box-Cox type transformations, it is tacitly assumed that the dose-response regression functions are linear, so that the intercepts and regression parameters provide all the necessary information for risk analysis. In a parallel-line assay the regression lines are assumed to have a common slope, while in the slope-ratio assay they are assumed to have a common intercept. A detailed statistical analysis of such simple semi-parametric and non-parametric bioassay models may be found in Sen (2000, 1997, 1984) among other places. Kim and Sen (2000) have considered a more general case where the two dose-response regression functions, denoted by $m_S(x)$ and $m_t(x)$ for the standard and the test preparation, are quite arbitrary. In parallel-line assay set ups, they are assumed to be parallel (without necessarily being linear); at the same time, some non-parametric tools are used to avoid the need to make specific distributional assumptions. The nearest neighbourhood methodology has been incorporated in the formulation of a robust estimation of the two regression surfaces, with a view to assess related risk properties. Their (Kim and Sen 2000) suggested robust (essentially large sample) procedure is suitable for EPT studies.

4.4. Meta Analysis

Let us finally consider the basic statistical problem of combining information from epimediologic and toxicologic studies. During the past two decades, there has been a steady growth of research literature on meta-analysis that aims to pool statistical evidence from independent studies. Basically, if there are similar EPT studies relating to comparable populations, we can adopt the meta-analysis techniques to draw more informative statistical conclusions. However, in the current case where we must incorporate the statistical evidence from dosimetric studies in the main theme of observational (epidemiologic) studies, there may be some compatability issues. In the context of hypothesis testing, the problem of pooling information may be simpler. There is a large body of statistical literature on the combination of independent tests of significance [mostly based on the *observed significance levels* (OSL) values]. Although such problems were studied in some way in the 1930s (Fisher 1934), the need for more elaborate analysis has grown considerably during the past two decades. Some attempts have also been made to extend the methodology to cover dependent tests as well. We refer to Mathew *et al.* (1993) and Sen (1999a, b) for some contemporary reviews. However, in the context of estimation of the EPT dose-response patterns or assessment of EPT risk, sufficient care should be made to ensure that pooling is valid. At the present time, there has been a steady flow of research methodology in environmental and clinical epidemiology which aims to provide satisfactory resolutions. *Matching, case-control* and *cohort* studies are all geared towards this. By nature, in clinical epidemiology, there is a greater emphasis on follow-up studies where some of the basic regularity conditions are different from standard situations. In this set up, *interim analysis* is quite common and in that context, combining statistical evidence from accumulating data sets (where independence and stationarity of associated stochastic processes may not hold) requires additional care. The well-known Simes method of combining P-values has been revisited for more complex situations and some extensions to accommodate dependent (accumulating) data models have been discussed in Sen (1999a, b). Most of these developments relate to hypothesis testing; in the estimation set up, the situation is more complex. There is a definite need for developing models for toxicologic studies more compatible with current epidemiologic studies. PBPK and mechanistic models are very appropriate in EPT studies; they relate the parameters in epidemiologic studies with those in toxicologic studies meaningfully, validating the pooling of statistical evidence from the two wings.

5. An Appraisal and Concluding Remarks

We illustrate some of the salient points discussed in earlier sections with the Iowa PGW Study (IPGW) (Hall *et al.* 2000). The IPGW Study was primarily an epidemiologic study. A sample of 3,695 from a population of 28,968 Iowans who served in the US military during the Persian (Kuwait/Iraq) Gulf War in the early 1990s was surveyed to assess a variety of health outcomes and exposures. A computer-assisted telephone interviewing system was administered. The sample was obtained using stratified random sampling with proportional allocation, with the stratification based on

(a) exposure status (whether or not deployed to the PGW military theatre),

(b) regular military versus guard/reserve status,

(c) branch of service,

(d) rank (enlisted versus officer),

(e) gender,

(f) race (white, black or other races), and

(g) age ($\leq$ 25 years or not).

There were some other covariates such as smoking (currently, previous, and not) which were not dichotomous. The distribution of the population, intended sample and achieved sample over these seven stratification variables are available in the data files that can be obtained on request from the principal investigators (Hall *et al.* 2000, IPGW Study 1997).

Of the seven stratification variables, the most important factor singled out was the exposure status: deployment to the PGW military theatre. Conceivably, the desert climate, exposure to UV-radiation and to chemical pollutants (more notably from the burning oil fields) and the tension prevailing at the battle lines were all hypothesized as potential factors for environmental health problems. With respect to this exposure end-point, the IPGW Study was a retrospective cohort type study (Hall *et al.* 2000). Subjects serving in the US military during the PGW were sampled on the basis of their exposure status. Data concerning subsequent health experiences were collected at the time of the survey administration (1995–96). However, because most of the questions in the IPGW Study survey related to current as well as recent health experiences rather than all post-PGW health experiences, it could not be classified as a cohort study. In that way, it combined elements of both the cohort and cross-sectional designs. As a result, for a majority of the health outcomes considered, the IPGW study allowed for the estimation of prevalence but not incidence. The selection-bias in this sampling design also made it difficult to apply standard statistical tools for the model specification and analysis schemes. Most common among the IPGW Study outcomes were dichotomous responses indicating the presence or absence of various adverse health conditions, including depression, bronchitis and cognitive dysfunction. Moreover, all these responses were based on the self-reported symptoms of the respondents. As a result, there was scope for misclassification and measurement errors.

In view of the study objectives and the nature of the outcome variables, Hall *et al.* (2000) considered the primary hypothesis H_1: The current health status of military personnel who were deployed to the PGW theatre (domains 1 and 3) is no different from that of military personnel serving at the time of the PGW who were not deployed to the PGW theatre (domain 2 and 4). They also considered three other hypotheses of secondary interest. The multiplicity of adverse health effect outcomes and the di(poly)-chotomous nature of the response variables made it difficult to use standard statistical tests for testing such hypotheses. For example, the use of conventional categorical (high-dimensional tables) models would have resulted in large degrees of freedom without a proportionately large non-centrality parameter and thereby could have very low power. Besides, not all cells would have adequately large frequencies to justify the conventional chi square approximation. Hall *et al.* (2000) bypassed some of these difficulties by using the Cochran-Mantel-Haenszel (CMH) procedure that leads to lower degrees of freedom and avoids the difficulties with low-count cells; their article (Jones *et al.* 1998) dealt with some methodologic issues. Ignoring the intricacies of the sampling design, Hall *et al.* (2000) assumed independence of the responses for any given health outcome which enabled them to make a direct use of the CMH procedure. However, in that way they could not force a statistical conclusion that the null hypothesis H_1 is not true. Probably, a more refined analysis based on a plausible dependence pattern would have improved the situation. More general methodology pertaining to polychotomous response variables in the set up of CMH procedures and multi-dimensional tables, discussed in Sen (1988), may be used to have more effective testing procedures. More important is the fact that deployment in the PGW theatre (domains 1 and 3), being a binary variable, by itself may not capture the full information. Perhaps a more informative picture could have been obtained by using logistic regression models in conjunction with duration and some measure of the distance with the line of action as covariates. Due to the fact that this study related to mostly the prevalence and not the incidence picture, there is some intrinsic loss of information. That possibly contributed to a lack of a significant difference in factor regarding deployment to the military theatre.

Hall *et al.* (2000) have addressed the sampling schemes very clearly, and have also raised the question of validity of conventional statistical analysis. As has been pointed out earlier, sacrificing the information on possible dependence (which is more likely the case) might have resulted in some distortion and loss of information, even when the CMH technique has been used Their attention to environmental epidemiology has been adequate and their dealing with categorical data models seems to be a correct step though due to the self-reporting nature of the health outcome variables, there is some scope for measurement errors and misspecifications. More complex models, discussed in Section 4, may perform better. There may be some concern with their simplistic modelling, though their efforts are praiseworthy. The main concern is the spatio-temporal dependence pattern. In what way domains 1 and 3 were more susceptible to adverse health outcomes than domains 2 and 4 may need more elaboration, and such epidemiologic-cum-environmental factors might affect the modelling to a certain extent. Further, with dichotomous response variables, to what extent would the conventional variogram-based statistical methodology be pertinent? Disregarding temporal considerations, typically, we have here multiple, correlated binary response variables along with a set of explanatory or auxiliary variables, some of which are also binary in nature. As such, it might be more relevant to formulate spatial statistical models for discrete and polychotomous response variables. This is a rather complex task and needs a lot of background material, some of which can be gathered from independent studies on related matters. The relevance of dosimetric studies is another important aspect. The toxicity and carcinogenecity of various environmental factors associated with the PGW deployment may have been studied by other groups, possibly with sub-human primates as subjects. In view of this, such information might be helpful in drawing statistical conclusions. On the whole, the efforts of the IPGW Study group in this objective assessment of adverse health effects should be commended. Undoubtedly, their continued studies will cast more light on this scientific assessment.

We may conclude with the following remarks. First, data collection relating to EPT sources and intensities needs statistical care. Secondly, identification of ET sources and impact should be assessed statistically with due support from environmental epidemiology, molecular biology and toxicology. The International Life Science Institute has been supporting research in EPT and has several monographs devoted to this general area. Various approaches related to assessing the inhalation toxicity of airborne toxicants along with pertinent methodological issues regarding dosimetric studies, interpretation of toxicokinetic studies and risk-assessment of RSPMs are discussed in Dungworth *et al.* (1988). Based on the developments on biomechanistic and PBPK models, as cited in earlier sections, structure-activity relationship information (Sen 2002) incorporation makes it more reasonable to combine the ET (statistical) information in the risk-assessment task in a meaningful way. Thirdly, for the assessment of EPT risk, observational studies as well as data banks from different demographic and health agencies constitute an indispensible source of statistical information that needs to be channelized in the formulation of risk measures. There are two important aspects of this assessment task. The first is, pooling statistical information from these diverse sources, which need not be independent or totally compatible in a statistical sense. Secondly, meta analysis (Hedges and Olkin 1985, Mathew *et al.* 1993) should be incorporated more aggresively. Most of these pooling procedures are devoted to combining independent tests. In the EPT context, there is a genuine need to combine information not only from similar studies conducted under compatible conditions (different regions or sectors) but also from different subject groups comprising sub-human primates (dosimetric studies) and human beings (clinical trials and observational studies).

At the present time, awareness in all these three sectors has been identified all over the world, though with considerable variation from developing to developed countries. Collection of epidemiologic data (Phase I) has gone through significant statistical uplift and more sophistications are indeed in evolution. There has also been an evolutionary growth of research literature on Phase II (dosimetric studies) though there is a need to extend the findings from rodents to human beings. Phase III (case studies) have also been advocated in many epidemiologic studies and they are potentially very valuable in the context of EPT risk analysis as well. Phase IV (clinical trials) is a comparatively recent undertaking. There should be enough guidelines to promote the use of epidemiologic studies involving human subjects without violating medical ethics, that is, entailing either significant side-effects of drugs or other risks due to them. The emergence of clinical trials, particularly in the Western Hemisphere, by government agencies as well as pharmaceutical companies (seeking governmental permission to release drugs in the market) has been received with mixed reactions. The statistical complexities underlying the modelling, data collection and data monitoring and statistical analysis of clinical trial datasets need to be appraised carefully so that the conclusions can be applied validly to the target population. Though the prospects are excellent, our (statisticians') task is by no means simple and routine. We must take up the challenge in the right perspective and provide satisfactory resolutions. In doing so, we need to work in collaboration with environmental scientists in the assessment of EPT, with epidemiologists to interpret the EPT risk and with all other researchers in a very broad interdisciplinary field. Only then we can achieve our overall goal.

REFERENCES

Abdelhardt, M., O. Jensen Moller, and H. Sand Hansen, (1985), 'Cancer of the Larynx, Pharynx and Oesophagus in Relation to Alcohol and Tobacco Consumption among Danish Brewery Workers', ***Danish Medical Bulletin***, 32, pp. 119-32.

Acheson, E.D. and M.J. Gardiner (1980), 'Asbestos: Scientific Basis for Environmental Control of Fibres', in J.C. Wagner (ed.), ***Biological Effects of Mineral Fibres***, IARC Scientific Publication No. 30, Lyon: International Agency for Research on Cancer.

Armitage, P. and R. Doll (1957), 'A Two-stage Theory of Carcinogenesis in Relation to the Age Distribution of Human Cancer', ***British Journal of Cancer***, 11, pp. 161-69.

Armitage, P. and R. Doll (1954), 'The Age Distribution of Cancer and a Multistage Theory of Carcinogenesis', ***British Journal of Cancer***, 8, pp. 1-12.

Barnett, V. and K.F. Turkman (eds.) (1997), ***Statistics for the Environment 3: Pollution Assessment and Control***, New York: John Wiley.

Barnett, V. and K.F. Turkman (eds.) (1994), ***Statistics for the Environment 2: Water Related Issues***, New York: John Wiley.

Barnett, V. and K.F. Turkman (eds.) (1993), ***Statistics for the Environment***, New York: John Wiley.

Barrett, J.C. and R.W. Wiseman (1987), 'Cellular and Molecular Mechanisms of Multistep Carcinogenesis: Relevance to Carcinogen Risk Assessment', ***Environmental Health Perspectives***, 76, pp. 65-70.

Bernardinelli, L., D. Clayton and C. Montomoli (1995a), 'Bayesian Estimates of Disease Maps: How Important are Priors?', ***Statistics in Medicine***, 14, pp. 2411-31.

Bernardinelli, L., D. Clayton, C. Pascutto, C. Montomoli, M. Ghistendi and M. Songini (1995b), 'Bayesian Analysis of Space-time Variation in Disease Risk', ***Statistics in Medicine***, 14, pp. 2433-43.

Bernardinelli, L. and C. Montomoli (1992), 'Empirical Bayes Versus Fully Bayes Analysis of Geographical Variation in Disease Risk', ***Statistics in Medicine***, 11, pp. 983-1007.

Bernardinelli, L., C. Pascutto, N.G. Best and W.R. Gilks (1997), 'Disease Mapping with Errors in Covariates', ***Statististics in Medicine***, 16, pp. 741-52.

Berry, G. and J.C. Wagner (1969), 'The Application of a Mathematical Model Describing the Times of Occurrence of Mesotheliomas in Rats following Inocculation with Asbestos', ***British Journal of Cancer***, 23, pp. 582-6.

Besag, J.E. (1989), 'Digital Image Processing Towards Bayesian Image Analysis', ***Journal of Applied Statistics***, 16, pp. 395-407.

Besag, J.E. (1974), 'Spatial Interaction and the Statistical Analysis of Lattice System (with discussion)', ***Journal of the Royal Statistical Society***, Series B, 36, pp. 192-236.

Besag, J.E., J. York and A. Mollie (1991), 'Bayesian Image Restoration, with Two Applications in Spatial Statistics (with discussion)', *Annals of the Institute of Statistical Mathematics*, 43, pp. 1-59.

Blot, W.J., L.E. Morris, R. Stroube, I. Tagnon, and J.F. Fraumeni (1980), 'Lung and Laryngeal Cancers in Relation to Shipyard Employment in Coastal Virginia', *Journal of the National Cancer Institute*, 65, pp. 571-5.

Breslow, N.E. and D. Clayton (1993), 'Approximate Inference in Generalized Linear Mixed Models', *Journal of the American Statistical Association*, 88, pp. 9-25.

Breslow, N.E. and N.E. Day (1987), *Statistical Methods in Cancer Research, Volume 2: The Design and Analysis of Cohort Studies*, International Agency for Research on Cancer, Scientific Publication, No. 82, Lyon : France.

Breslow, N.E. and N.E. Day (1980), *Statistical Methods in Cancer Research, Volume 1: The Analysis of Case-Control Studies*, IARC Scientific Publication No. 32, Lyon : International Agency for Research on Cancer.

Carlin, B.P. and T.A. Louis (1996), *Bayes and Empirical Bayes Methods for Data Analysis*, London: Chapman and Hall.

Carrat, F. and A.J. Valleron (1992), 'Epidemiological Mapping using the 'Kriging' Method: Application to Influenza -like Illness Epidemic in France', *American Journal of Epidemiology*, 135, pp. 1293-1300.

Case, R.A.M., M.E. Hosker, D.B. McDonald and J.T. Pearson (1954), 'Tumours of the Urinary Bladder in Workmen Engaged in the Manufacture and Use of Certain Dyestuff Intermediates in the British Chemical Industry', Part I: The Role of Aniline, Benzidine, Alpha-Naphthylamine and Beta-Naphthylamine', *British Journal of Industrial Medicine*, 11, pp. 75-104.

Chen-Mok, M. and P.K. Sen (1999), 'Nondifferentiable Errors in Beta-compliance Integrated Logistic Models', *Communications in Statistics, Theory and Methods*, 28, pp. 931-46.

Clayton, D. and L. Bernardinelli (1992), 'Bayesian Methods for Mapping Disease Risk', in P. Elliot, J. Cuzick, D. English and R. Stern (eds.), *Geographical and Environmental Epidemiology: Methods for Small Area Studies*, London: Oxford University Press.

Clegg, L.X., J. Cai and P.K. Sen (1999), 'A Marginal Mixed Baseline Hazards Model for Multivariate Failure Time Data', *Biometrics*, 55, pp. 805-12.

Cothern, C.R. and N.P. Ross (1994), *Environmental Statistics, Assessment and Forecasting*, Boca Raton: Lewis Publishers.

Cox, D.R. (1972), 'Regression Models and Life Tables (with discussion)', *Journal of the Royal Statistical Society*, Series B, 34, pp. 187-220.

Cox, L.H., D. Nychka and W.W. Piegorsch (eds.) (1998), *Case Studies in Environmental Statistics*, Lecture Notes, New York: Springer-Verlag.

Cressie, N.A.C. (1993), *Statistics for Spatial Data*, New York: John Wiley.

Datta, G., M. Ghosh and L.A. Waller (2000), 'Hierarchical and Empirical Bayes Methods for Environmental Risk Assessment', in P.K. Sen and C.R. Rao (eds.), *Handbook of Statistics*, 18, *Bioenvironmental and Public Health Statistics*, Amsterdam: North Holland.

Dey, D., P. Muller and D. Sinha (eds.) (1998), *Practical Nonparametric and Semi-parametric Bayesian Statistics*, New York: Springer-Verlag.

Diggle, P.J. (1990), *Time Series: A Biostatistical Introduction*, Oxford: Oxford Science Publication.

Diggle, P.J., K.Y. Liang and S.L. Zegar (1997), *Analysis of Longitudinal Data*, Oxford: Oxford Science Publication.

Doll, R. and R. Peto (1978), 'Cigarette Smoking and Bronchial Carcinoma : Dose and Time Relationships among Regular Smokers and Life-long Nonsmokers', *Journal of Epidemiology and Community Health*, 32, pp. 303-13.

Doll, R., R.E.W. Fisher, E.J. Gammon, W. Gunn, G.O. Hughes, F.H. Tryer and W. Wilson (1965), 'Mortality of the Gasworkers with Special Reference to Cancers of the Lung and Bladder, Chronic Bronchitis.. and Pneumoconiosis', *British Journal of Industrial Medicine*, 22, pp. 1-12.

Doll, R., L.G. Morgan and F. Speizer (1970), 'Cancer of the Lung and Nasal Sinuses in Nickel Workers', *British Journal of Cancer*, 24, pp. 623-32.

Dungworth, D., Kimmerle, G., Lewkowski, J., McClellan, R. and Stober, W. (eds.) (1988), *Inhalation Toxicology: The Design and Interpretation of Inhalation Studies and their Use in Risk Assessment*, New York: Springer-Verlag.

El-Masri, H.A. and C.J. Portier (1999), 'Replication Potential of Cells via the Protein Kinase C-MAPK Pathway : Application of a Mathematical Model', *Bulletin of Mathematical Biology*, 61, pp. 379-98.

Environmental Protection Agency (1994), 'Guidelines for Reproductive Toxicity Risk Assessment', US Environmental Protection Agency, Office of Research and Development, Washington, D.C.

Finney, D. (1964), *Statistical Methods in Biological Assay*, London: Griffin.

Fisher, R.A. (1934), *Statistical Methods for Research Workers*, Edinburg: Oliver and Boyd.

Fuller, W. (1986), *Measurement Error Models*, New York: John Wiley.

Gelfand, A.E. and A.F.M. Smith, (1990), 'Sampling Based Approaches to Calculating Marginal Densities', *Journal of the American Statistical Association*, 85, pp. 398-409.

Gelman, A., J.B. Carlin, H. Stern, and D.B. Rubin (1995), *Bayesian Data Analysis*, New York: Chapman and Hall.

Geman, S. and D. Geman (1984), 'Stochastic Relaxation, Gibbs Distribution and the Bayesian Restoration of Images', *IEEE Transactions on Pattern Analysis and Machine Intelligence*, 6, pp. 721-41.

Ghosh, M., K. Natarajan, A.W. Waller and Kim (1999), 'Hierarchical Bayes GLMs for the Analysis of Spatial Data: An Application to Disease Mapping', *Journal of Statistical Planning and Inference*, 75, pp. 305-18.

Ghosh, M. and J.N.K. Rao (1994), 'Small Area Estimation: An Appraisal (with discussion)', *Statistical Science*, 9, pp. 55-93.

Guttorp, P. and P.D. Sampson (1994), 'Methods of Estimating Heterogeneous Spatial Covariance Functions with Environmental Applications', in G.P. Patil and C.R. Rao (eds.) *Handbook of Statistics, Volume 12: Environmental Statistics*, Amsterdam: Elsevier.

Haenszel, W., D. Loveland and M.G. Sirken (1962), 'Lung Cancer Mortality as Related to Residence and Smoking Histories', *Journal of the National Cancer Institute*, 28, pp. 947-1001.

Hall, D.B., R.F. Woolson, W.R. Clarke and M.F. Jones (2000), 'Cochran-Mantel-Haenszel Techniques: Applications Involving Epidemiologic Survey Data', in P.K. Sen and C.R. Rao (eds.), *Handbook of Statistics, Volume 18: Bioenvironmental and Public Health Statistics*, Amsterdam: Elsevier.

Hallenbeck, W.H. (1993), *Quantitative Risk Assessment for Environmental and Occupational Health*, Boca Raton: CRC Press.

Hastings, W.K. (1970), 'Monte Carlo Sampling Methods Using Markov Chains and their Applications', *Biometrika*, 57, pp. 97-109.

Hedges, L.V. and Olkin, I. (1985), *Statistical Methods for Meta Analysis*, New York: Academic Press.

Hewitt, C.N. (1992), *Methods of Environmental Data Analysis*, Amsterdam: Elsevier.

Hirayama, T. (1975), 'Smoking and Cancer: A Prospective Study on Cancer Epidemiology Based on a Census Population in Japan', in *Proccedings of Third World Confererence on Smoking and Health*, Vol. 2, Washington D.C. : The US Health, Education and Welfare Department.

Hobbs, M.S.T., S. Woodward, B. Murphy, A.W. Musk and J.E. Elder (1980), 'The Incidence of Pneumoconiosis, Mesothelioma and other Respiratory Cancers in Men Engaged in Mining and Milling Crocidolite in Western Australia', in J.C. Wagner (ed.), *Biological Effects of Mineral Fibres*, IARC Scientific Publications No. 30, Lyon: International Agency for Research on Cancer.

Hougaard, P. (1998), 'Frailty', in P. Armitage and T. Cotton (eds.), *Encyclopedia of Biostatistics*, New York: John Wiley.

Infantine, P.F., R.A. Rinsky, J.K. Wagoner and R.J. Young (1977), 'Leukaemia in Benzene Workers', *Lancet* ii, pp. 76-78.

International Agency for Research on Cancer (1982), 'IARC Monographs on the Evaluation of the Carcinogenic Risk of Chemicals to Humans : Some Industrial Chemicals and Dyestuffs', IARC Scientific Publication No. 29, Lyon: International Agency for Research on Cancer.

Jones, M.F,. B.N. Doebbeling, D.B. Hall, T.L. Snyders, D.H. Barrett, A. Williams, K.H. Falter, J.K. Torner, L.F. Burmeister, R.F. Woolson, J.A. Merchant and D.A. Schwartz (1998), 'Methodologic Issues in a Population-based Health Survey of Gulf War Veterans', *Preventive Medicine and Environmental Health*, University of IOWA Technical Report No. 98-1.

Kaldor, J.M., J. Peto, D. Easton, R. Doll, C. Hermon and L. Morgan (1986), 'Models for Respiratory Cancer in Nickel Refinery Workers', *Journal of the National Cancer Institute*, 77, pp. 841-8.

Kim, H. and P.K. Sen (2000), 'Robust Procedures for Bioassyas and Bioequivalence Studies', *Sankhyā*, Series B, 62, pp. 119-33.

Kohn, M., G. Lucier, G. Clark, C. Sewall, A. Tritscher and C. Portier (1993), 'A Mechanistic Model of the Effects of Dioxin on Gene Expression in the Rat Liver', *Toxicology and Applied Pharmacology*, 120, pp. 138-54.

Kopp-Schneider, A., C. Portier and P. Bannasch (1998), 'A Model for Hepatocarcinogenesis Treating Phenotypical Changes in Focal Hepatocellular Lesions as Epigenetic events', *Mathematical Biosciences*, 148, pp. 181-204.

Lawson, A.B. and N. Cressie (2000), 'Spatial Statistical Methods for Environmental Epidemiology', in P.K. Sen and C.R. Rao (eds.), *Handbook of Statistics: Volume 18: Bioenvironmental and Public Health Statistics*, Amsterdam: Elsevier.

Lawson, A.B. and L. Waller (1996), 'A Review of Point Pattern Methods for Spatial Modelling of Events around Sources of Pollution', *Environmetrics*, 43, pp. 471-88.

Le, N.D. and J.V. Zidek (1992), 'Interpolation with Uncertain Spatial Covariance: A Bayesian Alternative to Kriging', *Journal of Multivariate Analysis*, 43, pp. 351-74.

Lee-Feldstein, A. (1983), 'Arsenic and Respiratory Cancer in Humans: Follow-up of Copper Smelter Employees in Montana', *Journal of the National Cancer Institute*, 70, pp. 601-10.

McCullagh, P. and Nelder, J.A. (1989), *Generalized Linear Models*, 2nd Ed., London: Chapman and Hall.

McIntyre, G.A. (1952), 'A Method for Unbiased Selection Sampling using Ranked Sets', *Australian Journal of Agricultural Research*, 3, pp. 385-90.

Marshall, R. (1991), 'Mapping Disease and Mortality Rates using Empirical Bayes Estimators', *Applied Statistics*, 40, pp. 283-94.

Mathéron, G. (1973), 'The Intrinsic Random Functions and their Applications', *Advances in Applied Probability*, 5, pp. 439-68.

Mathew, T., B.K. Sinha and L. Zhou (1993), 'Some Statistical Procedures for Combining Independent Tests', *Journal of the American Statistical Association*, 88, pp. 912-19.

Mazumdar, S., Y. Xu, D.R. Mattison, N.B. Sussman and V.C. Arena (2000), 'Statistical Methods for Reproductive Risk Assessment', in P.K. Sen and C.R. Rao (eds.), *Handbook of Statistics, Volume 18: Bioenvironmental and Public Health Statistics*, Amsterdam: Elsevier.

Moolgavkar, S. and A. Knudson (1981), 'Mutation and Cancer: A Model for Human Carcinogenesis', *Journal of the National Cancer Institute*, 66, pp. 1037-52.

Moolgavkar, S. and G. Luebeck, (1992), 'Multistage Carcinogenesis: A Population Perspective on Colon Cancer', *Journal of the National Cancer Institute*, 84, pp. 610-18.

Meng, Z., D. Dabdub and J.H. Seinfeld (1997), 'Chemical Coupling Between Atmospheric Ozone and Particulate Matter', *Science*, 2777, pp. 116-9.

Murphy, S.A. (1994), 'Consistency in a Proportional Hazards Model Incorporating a Random Effect', *Annals of Statistics*, 22, pp. 712-31.

Muttlak, H.A and L.L. McDonald (1990), 'Rank Set Sampling with Size Based Probability of Selection', *Biometrics*, 46, pp. 435-45.

Neyman, J. and E. Scott (1967), 'Statistical Aspects of the Problem of Carcinogensis', in L. LeCam and J. Neyman (eds.), *Fifth Berkeley Symposium on Mathematical Statistics and Probability*, Volume 2, Los Angeles: University of California Press.

Ott, W.R. (1995), *Environmental Statistics and Data Analysis*, Boca Raton: CRC Press.

Patil G.P. and C.R. Rao, (eds.) (1994), *Handbook of Statistics Volume 12: Environmental Statistics*, Amsterdam: Elsevier.

Patil, G.P.and C.R. Rao (1978), 'Weighted Distributions and Sized-bias Sampling with Applications to Wildlife Populations and Human Families', *Biometrics*, 34, pp. 179-89.

Patil, G.P., A.K. Sinha and C. Taillie (1992), 'Ranked Based Sampling in the Presence of a Trend on a Site', Pennsylvania State University Technical Report.

Person, J.C.G. and A. Turton (1993), *Statistical Methods in Environmental Health*, New York: Chapman and Hall.

Peto, J., H. Seidman and I.J. Selikoff (1982), 'Mesothelioma Mortality in Asbestos Workers: Implications for Models of Carcinogenesis and Risk Assessment', *British Journal of Cancer*, 45, pp. 124-35.

Piegorsch, W.W. (1994), 'Environmental Biometry: Assessing Impacts of Environmental Stimuli via Animal and Microbial Laboratory Studies', in G.P. Patil and C.R. Rao (eds.), *Handbook of Statistics, Volume 12; Environmental Statistics*, Amsterdam: Elsevier.

Piegorsch, W.W. and A.J. Bailer (1997), *Statistics for Environmental Biology and Toxicology*, London: Chapman and Hall.

Piegorsch, W.W. and A.J. Bailer (1994), 'Empirical Bayes Calculations of Concordance between Endpoints in Environmental Toxicity Experiments', *Environmental and Ecological Statistics*, 1, pp. 153-64.

Piegorsch, W.W. and L.H. Cox (1996), 'Combining Environmental Information 11: Environmental Epidemiology and Toxicology', *Environmetrics*, 7, pp. 309-24.

Piegorsch, W.W., E.P. Smith, D. Edwards and R.L. Smith (1997), 'Statistical Advances in Environmental Science', National Institute of Statistical Science Technical Report, No. 73, Research Triangle Park: North Carolina, USA.

Pope, C.A., J. Schwartz and M. Ranson (1992), 'Daily Mortality and PM10 Pollution in Utah Valley', *Archives of Environmental Health*, 42, pp. 211-7.

Portier, C. (1987), 'Statistical Properties of a Two-stage Model of Carcinogenesis', *Environmental Health Perspectives*, 76, pp. 125-32.

Portier, C. and A. Kopp-Schneider (1991), 'A Multistage Model of Carcinogenesis Incorporating DNA Damage and Repair', *Risk Analysis*, 11, pp. 535-43.

Portier, C. and C. Sherman (1994), 'The Potential Effects of Chemical Mixtures on the Carcinogenic Process within the Context of the Multistage Model', in R. Young (ed.), *Risk Assessment of Chemical Mixtures: Biological and Toxicological Issues*, New York: Academic Press.

Redmond, C.K. (1991), 'Introduction: Methods for Environmental Risk Assessment', *Environmental Health Perspectives*, 90, pp. 157.

Rosenblatt, M. (1999), *Gaussian and non-Gaussian Linear Time-series and Random Fields*, New York: Springer-Verlag.

Roth, H.D. and Y. Li (1996), 'Analysis of the Association between Air Pollutants with Mortality and Hospital Admissions in Burmingham', *Alabama, 1986-1990*, Technical Report, Roth Associate: Rockville, USA.

Sacks, J. and L.J. Steinberg (1994), 'Environmental Equity: Statistical Issues. Report of a Forum', *National Institute of Statistical Science Technical Report*, No. 11, Research Triangle Park: North Carolina, USA.

Samet, J.M., S.L. Zeger and K. Berhane (1995), 'The Association of Mortality and Particulate Air Pollution', in *Particulate Air Pollution and Daily Mortality: Replication and Validation of Selected Studies, The Phase I Report of the Particle Epidemiology Evaluation Project*, Health Effects Institute, Cambridge, Massachusetts.

Samet, J.M., S.L. Zeger, J.E. Kelsall, J. Xu and L.S. Kalkstein (1997), 'Air Pollution, Weather and Mortality in Philadelphia, 1973-1988', in *Particulate Air Pollution and Daily Mortality: Analyses of the Effects of Weather and Multiple Air Pollutants, The Phase IB Report of the Particle Epidemiology Evaluation Project*, Health Effects Institute, Cambridge, Massachusetts.

Sampson, P.D. and P. Guttorp (1992), 'Non-parametric Estimation of Nonstationary Spatial Covariance Structure', *Journal of the American Statistical Association*, 87, pp. 108-19.

Schwartz, J. (1993), 'Air Pollution and Daily Mortality in Birmingham, Alabama', *American Journal of Epidemiology*, 137, pp. 1136-47.

Schwartz, J. and D.M. Dockery (1992), 'Increased Mortality in Philadelphia Associated with Daily Air Pollution Concentrations', *American Review of Respiratory Diseases*, 145, pp. 600-4.

Schwartz, J. and A. Marcus (1990), 'Mortality and Air Pollution in London : A Time Series Analysis', *American Journal of Epidemiology*, 131, pp. 185-94.

Selikoff, I.J., E.C. Hammond and H. Seidman (1980), 'Latency of Asbestos Disease among Insulation Workers in the United States and Canada', *Cancer*, 46, pp. 2736-40.

Sen, P.K. (2002), 'Structure-activity Relationship Information Incorporation in Health Related Environmental Risk Assessment', *Environmetrics*, 13 (in press).

Sen, P.K. (2001a), 'Toxicology: Statistical Perspectives', *Current Science*, 80, pp. 101-9.

Sen, P.K. (2001b), 'Absorption and Ingestion Toxicology', in A.H. El-Shaarawi and W.W. Piegorsch (eds.), *Encyclopedia of Environmentrics*, UK: Wiley.

Sen, P.K. (2001c), 'Inhalation Toxicology', in A.H. El-Shaarawi and W.W. Piegorsch (eds.), *Encyclopedia of Environmetrics*, UK: Wiley.

Sen, P.K. (2000a), 'Bioenvironment and Public Health : Statistical Perspectives', in P.K. Sen and C.R. Rao (eds.), *Handbook of Statistics, Volume 18 : Bioenvironmental and Public Health Statistics*, Amsterdam : Elsevier.

Sen, P.K. (2000b), 'Non-parametrics in Bioenvironmental and Public Health Statistics', in P.K. Sen and C.R. Rao (eds.), ***Handbook of Statistics, Volume 18: Bioenvironmental and Public Health Statistics***, Amsterdam : Elsevier.

Sen, P.K. (1999a), 'Multiple Comparisons in Interim Analysis', ***Journal of Statistical Planning and Inference***, 82, pp. 5-23.

Sen, P.K. (1999b), 'Some Remarks on Simes Type Multiple Tests of Significance', ***Journal of Statistical Planning and Inference***, 82, pp. 139-45.

Sen, P.K. (1997), 'A Critical Analysis of Generalized Linear Models in Biostatistical Analysis', ***Journal of Applied Statistical Science***, 5, pp. 69-83.

Sen, P.K. (1988), 'Combination of Statistical Tests for Multivariate Hypotheses against Restricted Alternatives', in S. Dasgupta and J.K. Ghosh (eds.), ***Advances in Multivariate Statistical Analysis***, Calcutta: Indian Statistical Institute.

Sen, P.K. (1984), 'Nonparametric Procedures for some Miscellaneous Problems', in P.R. Krishnaiah and P.K. Sen (eds.), ***Handbook of Statistics, Volume 4: Nonparametric Methods***, Amsterdam : Elsevier.

Sen, P.K. and B.H. Margolin (1995), 'Inhalation Toxicology: Awareness, Indenfiability, Statistical Perspectives and Risk Assessment', ***Sankhyā***, Series B, 57, pp. 252-76.

Sen, P.K. and C.R. Rao (eds.) (2000), ***Handbook of Statistics, Volume 18 : Bioenvironmental and Public Health Statistics***, Amsterdam: Elsevier.

Sherman, C.D., C.J. Portier and A. Kopp-Schneider (1994), 'Multistage Models of Carcinogenesis: An Approximation for the Size and Number Distribution of Late-stage Clones', ***Risk Analysis***, 14, pp. 1039-48.

Smith, R.L. (1989), 'Extreme Value Analysis of Environmental Time Series: An Application to Trend Detection in Ground-level Ozone', ***Statistical Science***, 4, pp. 367-93.

Smith, R.L., J.M. Davis, J. Sacks, P. Speckman and P. Styer (2000), 'Regression Models for Air Pollution and Daily Mortality: Analysis of Data from Birmingham, Alabama, ***Environmetrics***, 11, pp. 745-63.

Stein, M. (1999), ***Interpolation of Spatial Data***, New York: Springer-Verlag.

Stocks, P. (1953), 'A Study of the Age Curve for Cancer of the Stomach in Connection with the Theory of the Cancer Producing Mechanism', ***British Journal of Cancer***, 7, pp. 407-17.

Stone, R. (1988), 'Investigation of Excess Environmental Risks around Putative Sources: Statistical Problems and a Proposed Test', ***Statistics in Medicine***, 14, pp. 2323-34.

Styer, P., N. McMillan, F. Gao, J. Davis and J. Sacks (1995), 'The Effect of Outdoor Airborne Particulate Matter on Daily Death Counts', ***Environmental Health Perspectives***, 103, pp. 490-7.

The Iowa Persian Gulf Study Group (1997), 'Self-reported Illness and Health Status Among Gulf War Veterans: a Population Based Study', ***Journal of the American Medical Association***, 277, pp. 238-45.

Walden, A.T. and P. Guttorp (1992), ***Statistics in the Environmental and Health Sciences***, New York: Halsted Press.

Walker, N., M. Wyde and G. Lucier (1998), 'Role of Estrogen in the Mechanism of Hepatocarcinogenesis Induced by TCDD in Female Rat Liver', in *Dioxin '98: 18th Symposium on Halogenated Environmental Organic Pollutants*, Stockholm, Sweden.

Waller, L.A., B.P. Carlin, H. Xia and A. Gelfand (1997), 'Hierarchical Spatio-temporal Mapping of Disease Rates', *Journal of the American Statistical Association*, 92, pp. 607-17.

Wedderburn, R.W.M. (1976), 'On the Existence and Uniqueness of the Maximum Likelihood Estimates for Certain Generalized Linear Models', *Biometrika*, 63, pp. 27-32.

Welsh, K., I. Higgins, M. Oh and C. Burchfiel (1982), 'Arsenic Exposure, Smoking and Respiratory Cancer in Copper Smelter Workers', *Archives of Environmental Health*, 37, pp. 325-35.

Whittemore, A. and J. Keller (1978), 'Quantitative Theories of Carcinogenesis', *Society of Industrial and Applied Mathematics Review*, 20, 1-30.

Whittemore, A. and A. McMillan (1981), 'Lung Cancer Mortality among US Uranium Miners: A Reappraisal', *Journal of the National Cancer Institute*, 71, pp. 489-99.

Xia, H. and B.P. Carlin (1998), 'Spatio-temporal Models with Errors and Covariates', *Statistics in Medicine*, 17, pp. 2025-44.

Zeger, S.L. and M.R. Karim (1991), 'Generalized Linear Models with Random Effects : A Gibbs Sampling Approach', *Journal of the American Statistical Association*, 86, 79-86.

Zeger, S.L. and K.Y. Liang (1986), 'Longitudinal Data Analysis for Discrete and Continuous Outcomes', *Biometrics*, 42, 121-30.

Zia, H., B.P. Carlin and L.A. Waller (1997), 'Hierarchical Models for Mapping Ohio Lung Cancer Rates', *Environmetrics*, 8, pp. 107-20.

Zidek, J.V. (1997), 'Interpolating Air Pollution for Health Impact Assessment', in V. Barnett and K.F. Turkman (eds.), *Statistics for the Environment 3: Pollution Assessment and Control*, New York: John Wiley.

6

Mangrove Forest Ecology of Sundarbans

The Study of Change in Water, Soil and Plant Diversity

A.K.M. NAZRUL-ISLAM

The vegetation pattern of different ecological zones of the Sundarbans mangrove forest was evaluated. The Polyhaline zone and also to a lesser extent Mesohaline Zone of the mangrove forest showed formation of consociation; the Oligohaline zone has mixed plant community. The ecological conditions of all the zones with reference to edaphic features and water chemistry and their seasonal variations were determined and it was found that calcium was the dominant cation followed by magnesium. Salinity (conductivity) in water samples from selected locations of various rivers showed strong variation and was several times higher in the Oligohaline zone at the end of the winter season (March) than in the rainy season and was due to the lack of fresh water supply from upstream through the river Ganges. Diversity was measured by H, the Shannon-Wiener Index. Ecological diversity was measured based on rarefaction of the actual samples,

$$E(S) = \sum \left\{ 1 - \left[(N_n - N_i) / \binom{N}{n} \right] \right\}.$$

The diversity values showed correlation with the ecological conditions. The rarefaction methodology was compared with a number of diversity indexes using identical data and was found to be influenced

---

I gratefully acknowledge the grant from the University Grants Commission (UGC) of Bangladesh for supporting this work.

by sample size. The abundance of species ranked from most to least abundant (in geometric series) was also calculated as

$$n_i = \binom{N}{k} K(1-K)^{i-1}$$

from the data of quadrats of circular plots of 2m radius. The data were also analysed with the log series and the Q statistic to bring plant diversity to a sharper focus. The result indicated a strong diversity in the Sundarbans mangrove forest.

1. Introduction

The origin of the word 'mangrove' is uncertain and there are many theories relating to its etymological evolution. It was derived possibly from the old Malay *manggi-manggi*, used to denote a specific family of mangrove trees and still used today in eastern Indonesia to describe the genus *Avicennia* (Mastaller 1997).

Mastaller (1997) also wrote that members of this genus were named *el gurm* (or el q'urum) by the Arabs and it is assumed that the word 'mangrove' may have been derived from the term 'mang-gurm', which is a combined form of the old Malay and Arabic names. To denote a dense stand of tropical trees, the English word 'grove' is used.

Mangroves were familiar to the ancients (MacNae 1968). Scientific study began after the European colonization in the 16th and 17th centuries. H. van Rheede tot Darkenstein described the mangrove in *Hortus indicus malabaricus* (Rheede 1678-1703). The *Herbarium amboinense* of Georg Everhard Rumpf (Rhumpius 1741-1755) may be mentioned in particular. Mangroves grow in a wide range of salinity conditions (Cintron *et al.* 1978).

The Sundarbans mangrove forest of Bangladesh is one of the single largest tracts in the world and is situated in the south-west corner of the country (Figure 6.1; Agroecological Region 13; between $21^o30'$ and $22^o30'$ N latitude; and 89° and 90° E longitude) and west of Brahmaputra-Meghna delta. This forest is very rich in flora and fauna. The forest has total area of 571,508 hectares, of which river channels and other water courses consist of 169,908 hectares. Plant communities occupy approximately 401,600 hectares (Nazrul-Islam 1995, 1994). Its low-lying areas are occupied by a network of small creeks and their margins are occupied by *Nypa fruticans, Acanthus ilicifolius, Phragmites karka* and *Porteresia coarctata.* Its soil environment (physical and chemical properties), vegetation pattern in various ecological zones and species diversity have been described (Nazrul-Islam 1995, 1987, 1986). In the present paper, the past and

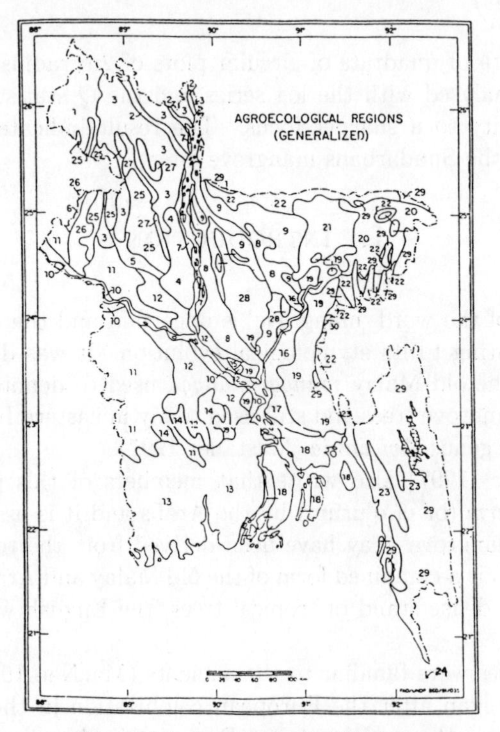

Figure 6.1: Agro-ecological Regions

1. Old Himalayan Piedmont Plain, 2. Active Tista Floodplain, 3. Tista Mander Floodplain, 4. Karatoya-Bangali Floodplain, 5. Lower Atrai Basin, 6. Lower Purnabhaba Floodplain, 7. Active Brahmaputra-Jamuna Floodplain, 8. Young Brahmaputra and Jamuna Floodplains, 9. Old Brahmaputra Floodplain, 10. Active Ganges Floodplain, 11. High Ganges River Floodplain, 12. Low Ganges River Floodplain, 13. Ganges Tidal Floodplain, 14. Gopalganj-Khulna Bils, 15. Arial Bil, 16. Middle Meghna River Floodplain, 17. Lower Meghna River Floodplain, 18. Young Meghna Estuarine Floodplain, 19. Old Meghna Estuarine Floodplain, 20. Eastern Surma-Kusiyara Floodplain, 21. Sylhet Basin, 22. Northern and Eastern Piedmont Plains, 23. Chittagong Coastal Plain, 24. St. Martin's Coral Island, 25. Level Barind Tract, 26. High Barind Tract, 27. North-Eastern Barind Tract, 28. Madhupur Tract, 29. Northern and Eastern Hills, 30. Akhaura Terrace

the present conditions of the Sundarbans mangrove forest and plant diversity are evaluated.

Works on Sundarbans' flora in the past have been reviewed by Prain (Reprinted 1979), based largely on work at the Indian Botanic Garden, Howrah. Roxburgh (Superintendent 1793-1814) received a number of interesting plants of the Sundarbans forest from William Carey and Buchanan-Hamilton. Roxburgh included these plants in *Hortus Bengalensis* (1814). Later, Falconer (Superintendent 1855-1861) collected plants from Sundarbans *(Cyomopsis psoraloides, Asphodelus tenuifolius)*. Thomson (Superintendent, 1845-1855) made an extensive collection from Sundarbans. Clarke (Officiating Superintendent 1869-1871) gave particular attention to the Sundarbans flora *(Merremia hederacea, Coldenia procumbens, Teramnus flexilis, Cladium riparium*, etc). Ellis (after 1880) sent an interesting plant *Oryza coarctata*. Heinig (1890) was posted as Deputy Conservator of Forests; his duties were to prepare a working plan of Sundarbans Forest Reserves. He collected plants in four or five seasons (1891-1894). His collections were sent to Calcutta Herbarium for identification where they were studied by G. King and D. Prain. C. B. Clarke provided an excellent account of the topography and vegetation of the Sundarbans. Clarke's work was presented in the form of a Presidential Address delivered at the anniversary meeting of the Linnean Society of London in 1895; he listed 60 species.

2. Topography of the Sundarbans

The Sundarbans (both Bangladeshi and Indian parts) form the southern part of the Gangetic delta between the Hughli river on the west and the Meghna river on the east. In Bangladesh, the western part of Sundarbans is demarcated by the Raimangal river (Figure 6.2). The area has a number of low-lying swampy islands formed by the main distributaries of the Ganges and their anabranches and connecting creeks. In the eastern section between the Madhumati (the Baleswar) and the Meghna, cultivation and clearings extend up to the sea-face. The central portion between the Baleswar and Raimangal rivers had reserved forest in the past; but at present the areas have decreased.

The various rivers within the Sundarbans and their courses are described below in a nutshell.

The Raimangal river separates the district of 24 Parganas from that of Khulna and also divides the protected forests from the reserved forests. This river has a course of 80 km. from Shaheb Khals to the sea.

The eastern parts of the Sundarbans are flushed with the water of the Ganges and its distributaries; but the western part of the Sundarbans resembles long arms of the sea rather than rivers and is subject to tidal influence. The water here is more saline than that of the central and eastern part of the Sundarbans. The character of vegetation is related salinity.

The Isamati river (Molingchu) begins near Halderkhali and after a course of 80 km through the Satkhira forest (the western half of central Sundarbans) joins the Barapunga river near the sea face. The Molingchu river is connected directly by the Firingi and various other canals.

The river Arpangasia, formed by the junction of the Kalpatta and the Kobadoc rivers near Burigoalini, flows southward for about 64 km between the Satkhira forest and the forests of the Khulna Reserves. This river is also called Barapunga in its lower reaches and is joined by Molingchu just before reaching the sea.

The Sibsa river originates at Deluti from the combination of a number of canals or streams derived from the Kobadoc on the west and from the Bhadra on the east. The Sibsa river system is connected with the Arpangasia by the Hansura, the Batlagang rivers and various other channels.

The Bhadra river leaves the Kobadoc river at Jhinargacha near Jessore, enters the Sundarbans forest reserves at the northern end of Sutarkhali and from this point has a course of 40 km before it is merged with the Sibsa.

The Passur river is an offshoot of the Bhrah at Khulna and flows 136 km to the sea. It is connected with the Bhadra by the Chunkori and the Bajna; with the Sibsa it is connected by the Chaila Bogi river. The Passur joins the Barashiala and the Sella rivers. A network of rivers and Khals connecting the tributaries of the Bangara river with Passur on the west and the Bhola river on the east can be noted.

The main channels in the Sundarbans are the Chachan Gang, the Andarmani, the Shellagang, the Aria Banki, the Pankassia, the Harintana and the Betmouri rivers.

The Bhola river beginning as a distinct river near Rampul, is connected on the north by means of a network of Khals with the Bhyrah and the Baleswar rivers. After flowing for 64 km, it joins the Pankassia near the junction of that river with the Haringhata river.

The Baleswar is a direct branch of the Ganges, the main stream of which it leaves near Pabna. In the upstream, this river is called the Gorai or Modhumati; it reaches the sea as Baleswar. From Bogi Khal southward it forms the eastern boundary of the reserve forests and separates the district of Khulna from that of Barisal. With the Bhola on the west, it is connected with the Jeodhara, the Chipa Bari, the Dansagar, the Sharankhola and the Sapala.

The rivers of the central and eastern Sundarbans are connected directly with the Ganges, and bring down an enormous volume of fresh water, especially during the monsoon. The nature and character of the vegetation in these areas are markedly affected by the river systems (Prain 1979, reprinted).

The heavy flow of water during the monsoon in the big rivers of the Sundarbans causes bank erosion. In contrast, the water flow in the winter season is virtually stopped because of the lack of water supply from the Ganges river.

3. Materials and Methods

Soil and water samples were collected from the various ecological zones (Figure 6.2) in different seasons. Soils were extracted for exchangeable cations (Gupta and Rorison 1974, Nazrul Islam and Rorison 1978). Calcium and magnesium were determined by atomic absorption spectrophotometry; sodium and potassium were analysed by the flame photometer; electrical conductivity was determined by a conductance bridge (manufactured by Griffin & Company); chloride and other ions in water samples were determined according to standard methods of the American Public Health Association (1977).

3.1. *Method of Quadrat Sampling*

At the outset, the sampling point was selected; three plots were chosen at the sampling site at a distance of 100 m to provide maximum habitat diversity. The sampling point was denoted as plot B; the plots A and C are located on either side of B at a distance of 100 m. At each plot a quadrat of radius 11.0 m was taken. All trees of various species of more than 20 cm diameter at breast height were counted and recorded.

4. Vegetation Pattern

4.1. Nature of the Vegetation

The vegetation of the Sundarbans can be considered in two ways:

(a) Vegetation of clearing spaces; and

(b) Vegetation of the forests proper.

4.1.1. Vegetation of Clearing Spaces
In the clearing spaces of the Oligohaline zone *Pandanus fascicularis, Flagellaria indica, Dalbergia spinosa, D. candenatensis, Clerodendron inerme* and *Achrostichum aureum* grow densely. In addition, in some places *Cyperus* sp., *Fimbristylis* sp. and *Paspaladum* sp. also invade the area.

4.1.2. Vegetation of the Forests Proper
The epiphytes found in this area include *Hoya parasitica* (rare). The mud flats covered at high tide are occupied by *Porteretia coarctata.* On the banks we find *Acanthis ilicifolius* and seedlings of various species and *Phragmites karka.* The top slope of the river bank is usually occupied by Keora *(Sonneratia apetala);* on the eastern side *Heritiera fomes* occupies major areas followed by *Excoecaria agallocha, Sonneratia apetala, Avicennia officinalis, Hibiscus tiliaceous, Cynometra ramiflora,* and *Nypa fruticans* which grow on the river bank. Inside the forest the open places are occupied by *Achrosticum aureum.* As one proceeds to the south (that is, towards the sea) the population of Sundri decreases and that of *Excoecaria agallocha* increases.

In the central portion of the forest *Excoecaria agallocha, Bruguiera sexangula, Xylocarpus mekongensis, Rhizophora apiculata, R. mucronata* and *Ceriops decandra* grow densely. On the western part *Aegiceras corniculatum,* and *Ceriops decandra* form consociation in many areas. Mixed communities are formed of *Bruguiera sexangula, Sonneratia apetala, Xylocarpus mekongensis* and *Kandelia candel.*

4.2. Plant Communities

The plant communities in the Sundarban mangrove forest vary in the three principal ecological zones, namely, Oligohaline, Mesohaline and Polyhaline. The following types of communities are dominant in these zones.

4.2.1. Oligohaline Zone

(i.) *Heritiera fomes - Cynometra ramiflora - Hibiscus tiliaceous*
(ii.) *Heritiera fomes - Avicennia alba*
(iii.) *Heritiera fomes - Excoecaria agallocha*
(iv.) *Rhizophora mucronata - Kandelia candel*
(v.) *Acrostichum aureum* in open places in highland only.

4.2.2. Mesohaline Zone

(i.) *Excoecaria agallocha - Heritiera fomes*
(ii.) *Sonneratia caseolaris - Tamarindus indica*
(iii.) *Sonneratia apetala - Nypa fruticans*
(iv.) *Porteretia coarctata - Phragmites karka* (in canal and river edges)
(v.) *Xylocarpus granatum - Bruguiera sexangula*

4.2.3. Polyhaline Zone

(i.) *Ceriops decandra* (forming consociation)
(ii.) *Aegiceras corniculatum* (forming consociation)
(iii.) *Ceriops decandra - Bruguiera sexangula*
(iv.) *Phoenix paludosa - Acrostichum aureum*
(v.) *Phoenix paludosa - Acanthus ilicifolius*

Amoora cucullata and Cerbera manghas were found on the river bank of the Oligohaline zone. Among climbers, *Derris trifoliata* (Kalilata) is abundant in the Oligohaline and Mesohaline zones; *Mucuna gigantea, Dendropthoe falcata* and *Caesalpinia cristata* are rare in the Oligohaline zone; *Finlaysonia maritima* is very rare in the Mesohaline zone. *Pandanus foetidus* grows on the river bank with *Heritiera fomes* and *Excoecaria agallocha* (matured seedlings). The woody parasite *Macrosolen cochinchinensis* (Pargassa) grows on crowns of trees in almost all the zones. *Ramalina calicaris* (lichen) is abundant on the branches of *Heritiera fomes* and *Excoecaria agallocha* in the Oligohaline and Mesohaline zones.

The dominant plant species of these ecological zones are given in Table 6.1. The Oligohaline zone has the highest number of species and the Polyhaline zone has the lowest.

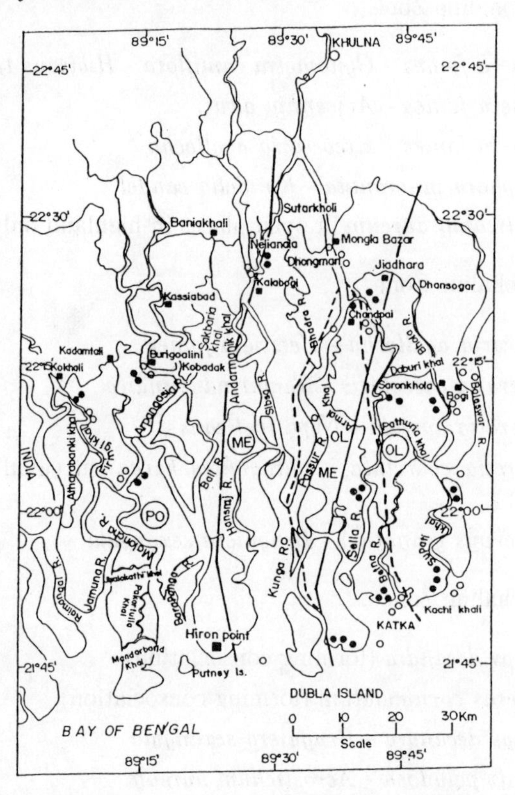

Figure 6.2: Map of Sundarbans Mangrove Forest showing various Rivers and locations from where Soil Samples (Solid Circles) and Water Samples (open Circles) were collected. Solid squares indicate the places where administrative offices are located within the forest. The whole forest is divided into three ecological zones (vertical lines) on the basis of salinity; OL = Oligohaline; ME = Mesohaline and PO = Polyhaline zones; OL & ME = Oligo-Mesohaline zone demarcated by dashed lines which becomes Oligohaline during the monsoon season when there is enough fresh water supply from the Ganges through Baleshwar river (extreme east) and becomes Mesohaline during winter when the water in the Ganges is diverted through the Farakka Barrage.

5. Water Chemistry and Soil Environment

Since the Sundarbans forest is flooded two times in a day by tides, the study of water chemistry of the Bay of Bengal was also considered (Table 6.2, Nazrul-Islam 1985). Water samples were collected from different locations in the Bay of Bengal. The pH varied from 7.9 to 8.3; conductivity was in the range of 24,000 (Karnafuli estuary) to 40,500 micromhos/cm; chloride varied from 9,926 to 16,990 mgl^{-1}. Dissolved oxygen and COD values were not alarming. Turbidity values indicate that water of the Bay of Bengal is less turbid compared to other seas.

Table 6.1

Dominant Vegetation of Three Ecological Zones

Oligohaline Zone	Mesohaline Zone	Polyhaline Zone
Excoecaria	*Excoecaria*	*Aegiceras*
agallocha (Ab)	*agallocha* (Ab)	*corniculatum* (Ab)
Heritiera	*Heritiera*	*Bruguiera*
fomes (Ab)	*fomes* (Ab)	*gymnorrhiza* (Ab)
Porteretia	*Rhizophora*	*Ceriops*
coarctata (Ab)	*apiculata* (Ab)	*decandra* (Ab)
	Rhizophora	
Acanthus	*mucronata* (Ab)	*Acanthus*
ilicifolius (F)	(shows top dying)	*ilicifolius* (F)
Amoora		*Acrostichum*
cucullata (F)	*Bruguiera*	*aureum* (F)
Avicennia	*sexangula* (F)	*Brownlowia*
alba (F)	*Ceriops*	*tersa* (F)
A. officinalis (F)	*decandra* (F)	*B. sexangula* (F)
Barringtonia	*Kandelia*	*Kandelia*
racemosa (F)	*candel* (F)	*candel* (F)
Derris	*Nypa*	*Nypa*
trifoliata (F)	*fruitcans* (F)	*fruticans* (F)
Hibiscus	*Sonneratia*	*Phoneix*
tiliaceous (F)	*apetala* (F)	*paludosa* (F)
Sonneratia		*Xylocarpus*
apetala (F)	*Dalbergia* (Oc)	*granatum* (F)
S. caseolaris (F)	*candenatensis* (Oc)	*X. mekongensis* (F)
	Dalbergia	
Acrostichum	*spinosa* (Oc)	*Macrosolen*
aureum (Oc)	*Porteretia*	*cochinchinensis* (R)
Cerbera	*coarctata* (Oc)	(Woody parasite)
manghas (Oc)	*Tamarix*	
Cynometra	*indica* (Oc)	
ramiflora (Oc)		

Oligohaline Zone	Mesohaline Zone	Polyhaline Zone
Lumnitzera racemosa (Oc)	*Brownlowia tersa* (R)	
Nypa fruticans (Oc)	*Pandanus foetidus* (R)	
Pandanus foetidus (Oc)	*Petunga roxburghii* (R)	
Pharagmites karka (Oc)	*Phoenix paludosa* (R)	
Pongamia pinnata (Oc)	*Sarcolobus globosus* (R)	
Sapium indicum (Oc)		
Tamarix indica (Oc)	*Clerodendron inerme* (VR)	
Typha elephantina (Oc)	*Finlaysonia maritima* (VR)	
Caesalpinia cristata (R)	*Mucuna gigantea* (VR)	
Clerodendron inerme (R)		
Dendropthoe falcata (R)		
Flagellaria indica (R)		
Ceriops decandra (VR)		
Mucuna gigantea (VR)		
Tetrastigma bracteolatum (VR)		

Ab = Abundant; F = Frequent; Oc = Occasional; R = Rare; VR = Very rare

Silt is the major componet of the soil of the Sundarbans; it is followed by clay. Clay content was generally higher in the polyhaline zone than in other zones (Nazrul-Islam 1995). Calcium was the dominant cation and was highest in the Oligohaline zone, followed by the Mesohaline and Polyhaline zones (Table 6.3); sodium was significantly higher (P=0.05) in the Polyhaline zone than in the other zones. Seasonal variation of soil salinity (conductivity) was observed (Table 6.4). During the monsoon season the large volume of fresh water coming from upstream through the Gorai river, a distributory of the Ganges, dilutes the soil salinity and hence the values are low. It is important to mention that in some places in the Polyhaline zone, soil salinity (conductivity, as measured in micromhos/cm) was more than 6,000.

Table 6.2

Analysis of Water Samples from the Bay of Bengal

Location N	Location E	*pH*	E.C. micromhos/cm	Chloride mg/l	Total Alkalinity mg/l	D.O. mg/l	Turbidity in JTU
(Collected in Jaunary 1981)							
$22^o 16'$	$91^o 45'$	8.2	46500	16990.0	250	5.9	30
$21^o 10'$	$91^o 15'$	8.1	40100	14950.0	160	6.4	25
$20^o 13'$	$92^o 15'$	8.3	40000	11120.0	160	5.8	25
$21^o 40'$	$89^o 42'$	8.1	35400	10660.0	220	5.6	25
$21^o 31'$	$89^o 30'$	8.1	40500	10870.0	180	6.0	25
$21^o 35'$	$90^o 10'$	8.0	30500	10250.0	180	6.0	25
(Collected in February 1982)							
							COD mg/l
Karnafuli River		7.9	2000	496.5	140		80
Estuary		8.0	24000	10980.0	190		45
$21^o 56'$	$91^o 40'$	8.1	25000	10425.0	250		46
$21^o 41'$	$91^o 37'$	8.2	24000	9926.0	220		42
$21^o 12'$	$90^o 15'$	8.1	36000	14592.0	260		41
$21^o 13'$	$89^o 46'$	8.0	40000	16478.0	210		40
$21^o 25'$	$91^o 15'$	8.0	25500	11316.0	210		40
$21^o 32'$	$91^o 31'$	8.1	28000	11210.0	260		40

Water samples of the Rupsha river (attached to Khulna Town) had a conductivity of 4,800 micromhos/cm in April which is 5-15 times higher than the values obtained in March and the monsoon season (June, July and November) (Table 6.5). In the winter months, the low discharge of river water into the Ganges clearly indicates the intrusion of sea water upto Khulna city. In the Polyhaline zone, water conductivity in the Arpangasia and Firingi rivers was measured at 26,500 and 27,000 micromhos/cm respectively. Upon a recent visit (1994) to these rivers, water conductivity of 40,000 micromhos/cm was noted.

6. Plant Diversity

The simplest measure of species diversity is to count the number of species and their heterogeneity. Peet (1974) suggested that the different concepts of number of species and relative abundance should be combined into a single concept of heterogeneity. According to Preston (1948), the X axis (number of individuals represented in sample) rests on a geometric (logarithmic) scale rather than on an arithmetic scale. After the conversion, the relative abundance data takes the form of a bell-shaped, normal distribution and is referred to as the log-normal distribution. The log-normal distribution is described by the formula

$$Y = Yo^{e^{-(aR)^2}}$$

where

Y = number of species to occur in the Rth octave to the right or left of the modal class

Yo = number of species in the modal octave (the largest class)

a = constant describing the amount of spread of the distribution

c = 2.71828 ... (a constant)

Table 6.3
Physico-Chemical Properties of Soils

Ecological Zones	*pH*	Organic Matter (%)	Na $\mu g/g$	K $\mu g/g$	Ca $\mu g/g$	Mg $\mu g/g$
Oligohaline	6.9	6.8	300-500	250-750	3,029-4,695	438-1,579
Mesohaline	7.0	6.5	575-800	250-375	2,612-3,003	749-939
Polyhaline	7.0	6.5	850-2,500	375-890	1,844-3,590	1,008-1,672

Table 6.4
Seasonal Variation of Soil Salinity, 1982-1983 (Micromhos/cm)

Ecological Zones	July	September	November	January	March
Oligohaline	425	450	450	600	650
Oligo-Mesohaline	2,200	2,200	2,050	2,250	2,250
Mesohaline	2,800	3,050	3,150	3,250	3,400
Polyhaline	3,250	3,500	3,450	4,800	6,850

Table 6.5
Seasonal Variation of Electrical Conductivity in Micromhos/cm in Water Samples of the Rivers of Sundarbans Mangrove Forest

Location	1983		1984		1985	
(River)	Mar	Jun	Jul	Nov	Jan	Apr
Khulna	1350	350	300	270	337	4800
(Rupsha)	(350)	(150)	(100)	(200)	(250)	(1800)
Mongla	9700	1050	708	300	2100	10500
(Passur)	(3750)	(550)	(410)	(250)	(1000)	(6200)
Chandpai	–	5900	831	490	5200	8700
(Sella)		(3150)	(425)	(700)	(2250)	(3800)
Sharankhola	–	860	338	315	3150	5700
(Bhola)		(450)	(105)	(450)	(1250)	(2500)
Bogi	–	280	215	260	530	3000
(Baleswar)	–	(140)	(100)	(350)	(400)	(1250)
Katka	–	–	9000	10750	14000	15500
(Bay)			(5500)	(3900)	(4500)	(10250)
Nalianala	22400	18500	4324	7800	8400	15800
(Sibsa)	(8750)	(5325)	(1950)	(2800)	(3100)	(11500)
(Bazbaza)	26500	20000	–	–	–	–
(Khal)	(10750)					
Burigoalini	26500	16000	12050	10150	12300	18000
(Arpangasia)	(11000)		(4200)	(4700)	(7200)	(10750)
(Firingi)	27000	–	–	–	–	25500
	(12500)					(12500)

'–' indicates data not available.
Data in parentheses are the amount of chloride in mg/l

The shape of the log-normal curve is supposed to be the characteristic for any particular community. Additional sampling of a community should move the log-normal curve to the right along the abscissa but not change its shape. Preston (1948) showed that for many cases, a = 0.2. If the number of species in the whole community is known, the entire equation for the log-normal curve can be specified. This implies that species diversity can be measured by counts of the number of species and that relative abundance follows quite specific rules in biological communities.

When the species abundance distribution is log-normal, the total number of species in the community can be calculated, including rare species not yet collected. This is done by extrapolating the bell-shaped curve below the class of minimal abundance and measuring the area (Figure 6.3a and Figure 6.3b). The formula for the total number of species (May 1975) is

$$S = Y_O\sqrt{\pi/a}$$

where

S	=	total number of species in the community
Yo	=	number of species in the modal octave
π	=	3.14159
a	=	a constant describing the spread of log-normal distribution (often equal to 0.2)

Heterogeneity of community was measured by using concepts of information theory (Peet 1974). One approach to species diversity involves measures of the heterogeneity of a community. Several measures of heterogeneity are in use (Peet 1974) and the most popular measure of heterogeneity has been borrowed from information theory. The main objective of information theory is to try to measure the amount of order (or disorder) contained in a system (Margalef 1958). This was measured by the Shannon-Wiener function (Krebs 1978).

$$H = -\sum_{i-1}^{S}(p_i)(log_2 p_i)$$

where

H	=	information content of sample (bits/individual)
S	=	number of species
p_i	=	Proportion of total sample belonging to ith species

Heterogeneity was measured and was described earlier (Nazrul-Islam 1994). Values of indices of plant diversity are given in Table 6.6 (Kempton and Taylor 1978). These values were compared with the deciduous sal (*Shorea robusta*) forest of Modhupur Tract.

Strictly speaking, the Shannon-Wiener function that measures information content should be used only on random samples drawn from a large community in which the total number of species present is known. Pielou (1966) discussed information measures appropriate to other circumstances. Two components of diversity are combined in the Shannon-Wiener function: the number of species and the equitability or evenness of allotment of individuals among the species (Lloyd and Ghelardi 1964). A greater number of species increases the species diversity, and a more even or equitable distribution among species will also increase species diversity as measured by the Shannon-Wiener function.

If the study area can be successfully delimited in space and time and the constituent species enumerated and identified, species richness provides an extremely useful measure of diversity. When a sample is obtained, it becomes necessary to distinguish between numerical species richness, that is the number of species per specified number of individuals or biomass (Kempton 1979), and species density which is the number of species per specified collection area (Hurlbert 1971) (number of species per m^2). However, it is not always possible to ensure that all sample sizes are equal and the number of species increases invariably with sample size. To cope with this problem, Sanders (1968) devised a technique called Rarefaction, for calculating the number of species expected in each sample if all samples were of a standard size. This method was modified by Hurlbert (1971) to produce an unbiased estimate and the formula is:

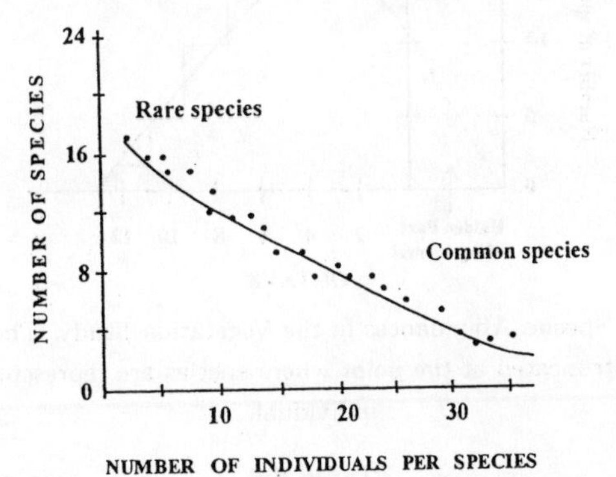

Figure 6.3a: Relative Abundance of Plant Species. There are 18 species represented by only a single specimen (Rare species).

$$E(S) = \sum_{i=1}^{S} \left[1 - \frac{\binom{N - n_i}{n}}{\binom{N}{n}} \right]$$

where

$E(S)$	=	the expected number of species in the rarefied sample
n	=	standardized sample size
N	=	the total number of individuals recorded in the sample to be rarefied
n_i	=	number of individuals in the ith species recorded in the sample to be rarefied

In the present study, species richness was calculated according to the above formula and is shown in Table 6.7. The simplest approach is to take the number of individuals in the smallest sample as the standardized sample size (n). Thirty three individuals were recorded in Area 'A' but only 17 were found in Area 'B'. How many species would one has expected in Area 'A' if, it too had contained 17 individuals? The answer is approximately 12, as shown in the calculations in Table 6.7 as 11.19.

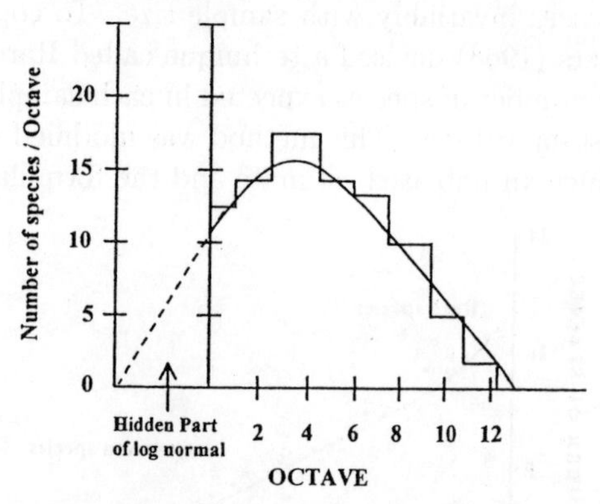

Figure 6.3b: Species Abundances in the Vegetation Study. The log-normal distribution is truncated at the point where species are represented by a single individual.

Table 6.6

Values of Indices of Plant Diversity for Mangrove and Decidous Forests.

Index	Formula	Mangrove	Deciduous
Simpson's (Inverse)	$\frac{1}{\frac{S}{\sum_i \rho_i^2}}$	81.2	30.2
Shannon-Wiener (Exponent)	$\exp\left(-\sum_i^S \rho_i \log_2 \rho_i\right)$	147.5	64.2
Q Statistic	$\frac{\frac{1}{2}S}{\log\frac{R_2}{R_1}}$	98.6	34.5
	Log-normal gamma	98.0	38.2
	log series ($\propto$)	101.0	27.2
Whittaker's	$\frac{S}{4\left[\left(\sum_i^S (\log \rho_i - \log \rho)^2\right)\right]}$	75.1	26.8
Log Cycle	$\frac{S}{\log \rho_i - \log \rho_S}$	78.2	20.3

The abundance of species ranked from most to least abundant (geometric series, May 1975) was also calculated as

$$n_i = NC_K K(1-K)^{i-1}$$

where

K = the proportion of available niche space that each species occupies,

n_i = the number of individuals in the ith species;

N = the total number of individuals;

$C_K = [1-(1-K)^S]^{-1}$;

and

K is a constant which ensures that $\sum n_i = N$. Ecological diversity was calculated (Table 6.8) according to Magurran (1988).

Table 6.7

Data of Quadrats (11 m Radius) in Two Areas of the Oligohaline Zone; Calculations for Area 'A' are Shown in the Middle Column (Each Value is the Mean of 10 Quadrats Taken at Random)

Species	Area 'A'	Calculation for Area 'A'	Calculation for Area 'B'
Heritiera fomes	7	1.00	4
Excoecaria agallocha	3	0.90	3
Avicennia officinalis	2	0.77	1
Sonneratia apetala	3	0.99	1
Petunga roxburghii	0	–	1
Nypa fruticans	2	0.77	1
Cynometra ramiflora	3	0.90	0
Pandanus foetidus	1	0.52	0
Phoenix paludosa	2	0.77	1
Cerbera manghas	1	0.52	0
Lumnitzera racemosa	1	0.52	0
Kandelia candel	1	0.52	0
Rhizophora mucronata	1	0.52	0
Amoora cucullata	1	0.52	1
Hibiscus tiliaceous	2	0.77	0
Dalbergia spinosa	0	–	1
Acrostichum aureum	2	0.77	1
Flagellaria indica	1	0.52	0
Barringtonia racemosa	0	–	1
Clerodendron inerme	0	–	1
Number of Species (S)	16		12
Number of individuals (N)	33		17
Expected number of species for Area 'A' E(S)		11.19	

1. In order to fit the geometric series it is necessary to begin by estimating the constant K. This is done by iterating the

following equation (May 1975).

$$\frac{N_{\min}}{N} = [K/(1-K)][(1-K)^S]/[1-(1-K)^S]$$

$N_{\min}$ is the number of individuals in the least abundant species. In this example

$$\frac{N_{\min}}{N} = 0.004694835$$

To solve the equation requires using successive values of K until the two sides of the equation balance.

For example:

If $K = 0.2$ then $\frac{N_{min}}{N} = [K/(1-K)] \times [(1-K)^s]/[1-(1-K)^S] = 0.003655$

If $K = 0.18$ then $N_{min}/N = 0.005176$

If $K = 0.19$ then $N_{min}/N = 0.004359$

If $K = 0.185$ then $N_{min}/N = 0.004753$

If $K = 0.1857$ then $N_{min}/N = 0.004694$

Table 6.8

Geometric Series Analysis from Data of Quadrats on Four 11 m Radius Circular Plots

Species	Observed	Expected	χ^2
Heritiera fomes	87	40.370	53.86
Hibiscus tiliaceous	33	32.870	0.11
Acrostichum aureum	24	26.770	0.29
Excoecaria agallocha	12	21.796	4.40
Cynometra ramiflora	9	17.749	4.31
Pandanus foetidus	9	14.454	2.06
Sarcolobus globosus	8	11.769	1.20
Amoora cucullata	6	9.581	1.34
Acanthus ilicifolius	4	7.804	1.85
Nypa fruticans	4	6.354	0.87
Ceriops decandra	3	5.175	0.91
Phoenix paludosa	3	4.214	0.35
Sonneratia apetala	3	3.432	0.05
Xylocarpus mekongensis	2	2.794	0.23
Bruguiera sexangula	2	2.275	0.03
Lumnitzera racemosa	1	1.853	0.39
Avicennia officinalis	1	1.508	0.17
Barringtonia racemosa	1	1.228	0.04
Rhizophora mucronata	1	1.000	0.00
	$\sum n_i = 213$	213	$\sum \chi^2 = 72.46$

Number of Species (S) = 19
Number of individuals (N) = 213

2. With K estimated as 0.1857 it is now possible to obtain the values of C_K

$$\begin{aligned} C_K &= [1-(1-K)^S]^{-1} \\ &= [1-(1-0.1857)^{19}]^{-1} \\ &= [1-(0.8143)^{19}]^{-1} \\ &= (1-0.020179214)^{-1} \\ &= (0.979820785)^{-1} \\ &= 1.020594802 \end{aligned}$$

The expected number of individuals for each of the 19 species was calculated. Thus, for the most abundant species

$$\begin{aligned} n_i &= NC_K K(1-K)^{i-1} \\ &= 213 \times 1.020594802 \times 0.1857 \times (1-0.1857)^0 \\ &= 40.37 \end{aligned}$$

The total number of individuals is N; and when the proportion of 'K' is taken from the total sample, the mathematical formula takes the form of equation (1); and from the remaining sample, when the proportion 'K' is taken upto (2), (3) and (4) and then upto the sth species, the mathematical equations take the following forms.

$$\begin{aligned} n_1 NK &= NK(1-K)^{1-1} && (1) \\ n_2\,(N-NK)K &= NK(1-K) \\ &= NK(1-K)^{2-1} && (2) \\ n_3[N-NK-NK(1-K)]K &= NK(1-K)^2 && (3) \\ &= NK(1-K)^{3-1} \\ n_4\,[N-NK-NK(1-K)-NK(1-K)^2]K &= NK(1-K)^3 && (4) \\ &= NK(1-K)^{4-1} \end{aligned}$$

For the sth species

$$\begin{aligned} n_s &= NK(1-K)^{S-1} \\ &= C_K NK(1-K)^{S-1} \\ \sum n_s &= N.C_K \end{aligned}$$

REFERENCES

American Public Health Association (1977), 'Standard Methods for the Examination of Water and Waste Water', Washington D.C.: American Public Health Association.

Cintron, G., A.E. Lugo, P.J. Pool and G. Morris (1978), 'Mangrove of Arid Environments in Puerto Rico and Adjacent Islands', *Biotropica*, 10, pp. 110-21.

Gupta, P.L. and I.H. Rorison (1974), 'Effects of Storage in the Soluble Phosphorus and Potassium Content of Some Derbyshire Soils', *Journal of Applied Ecology*, 11, pp. 1197-1207.

Hurlbert, S.H. (1971), 'The Non-concept of Species Diversity: A Critique and Alternative Parameters', *Ecology*, 52, pp. 577-86.

Kempton, R.A. (1979), 'Structure of Species Abundance and Measurement of Diversity', *Biometrics*, 35, pp. 307-22.

Kempton, R.A. and L.R. Taylor (1978), 'The *Q*-Statistic and the Diversity of Floras', *Nature*, 275, pp. 252-53.

Krebs, C.J. (1978), *Ecology: The Experimental Analysis of Distribution and Abundance*, Harper International Edition, New York: Harper & Row Publishers.

Lloyd, M. and R.J. Ghelardi (1964), 'A Table for Calculating the "Equitability" Component of Species Diversity', *Journal of Animal Ecology*, 33, pp. 217-25.

MacNae, W. (1968), 'A General Account of the Fauna and Flora of Mangrove Swamps and Forests in the Indo-West Pacific Region', *Advances in Marine Biology*, 6, pp. 73-270.

Magurran, A.E. (1988), *Ecologial Diversity and its Measurement*, London: Chapman and Hall.

Margalef, R. (1958), 'Information Theory in Ecology', *General Systematics*, 3, pp. pp. 36-71.

Mastaller, M. (1997), *Mangroves: The Forgotten Forest between Land and Sea*, Kuala Lumpur: Tropical Press.

May, R.M. (1975), 'Patterns of Species Abundance and Diversity' in M.L. Code and J.M. Diamokd (eds.), *Ecology and Evolution of Communities*, Cambridge, Massatussetts: Harvard University Press.

Nazrul-Islam, A.K.M. (1995), 'Ecological Conditions and Species Diversity in Sundarban Mangrove Forest Community, Bangladesh', in M.A. Khan and I.A. Unger (eds.), *Biology of Salt Tolerant Plants*, Michigan: Book Crafters.

Nazrul-Islam, A.K.M. (1994), 'Environment and the Vegetation of Sundarban Mangrove Forest', in H. Leith (ed.) *Towards Rational Use of High Salinity Tolerant Plants*, Vol. 1, Dordrecht: Kluwer Academic Publishers.

Nazrul-Islam, A.K.M. (1987), 'The Ecology of Sundarban Mangrove Forest', MAB Regional Training Workshop Proccedings on The Ecology and Conservation of the Tropical Humid Forests of the Ind-Malayan Realm, Sri Lanka.

Nazrul-Islam, A.K.M. (1986), 'Environment of Sundarban Mangrove Forest', Proceedings of the SAARC Seminar on Protection of Environment from Degradation, Dhaka.

Nazrul-Islam, A.K.M. (1985), 'The Ecology of Bay with Notes on Coastal Edaphic Features', *Journal of NOAMI*, 2, pp. 31-5.

Nazrul-Islam, A.K.M. and I.H. Rorison (1978), 'Field Investigation of Seasonal Oxidation-reduction Conditions in Soil and of Changes of Shoots of Associated Species', *Dacca University Studies*, 26, pp. 57-65.

Peet, R.K. (1974), 'The Measurement of Species Diversity', ***Annual Review of Ecological Systematics***, 5, pp. 285-307.

Pielou, E.C. (1966), 'The Measurement of Diversity in Different Types of Biological Collection', ***Journal of Theoretical Biology***, 13, pp. 131-44.

Prain, D. (1979, Reprinted), ***Flora of Sundarbans***, Delhi: Periodical Expert Book Agency.

Preston, F.W. (1948), 'The Commonness and Rarity of Species', ***Ecology***, 29, pp. 254-83.

Rheede (tot Drakenstein) H. 1678-1703. ***Hortus indicus malabaricus***, Amsterdam: Amstelodami.

Rumphius, G.E. 1741-55, ***Herbarium Amboinense***, Amsterdam: Amstelodami.

Sanders, H.L. (1968), 'Marine Benthic Diversity: A Comparative Study', ***American Naturalist***, 102, pp. 243-82.

Simpsons, E.H. (1949), 'Measurement of Diversity', ***Nature***, pp. 163-688.

Williams, C.B. (1946), 'Yules Characteristics and the "Index of diversity", ***Nature***, 157, pp. 482.

7

Classified Raster Map Analysis for Sustainable Environment and Development in the 21st Century

A Perspective

G. P. Patil

Geospatial data form the foundation of an information-based society. Cell-based raster data are well-suited to represent geographic phenomena and help model land cover and land use characteristics and relevant attributes. Information technologies capable of credible raster map analysis and change detection that is equipped with remote sensing and other geospatial data are needed for integrated regional assessment and development in the 21st century.

The paper discusses examples of ecosystem health assessment at landscape and watershed scales using relatively new concepts and measures of landscape fragmentation. The methods include hierarchical transition matrix analysis, fragmentation profile analysis, echelon analysis, and nested area sampling analysis.

Insightful discussions with and valuable input from G.D. Johnson, W.L. Myers, D.J. Rapport and C. Taillie are greatly appreciated. Prepared with partial support from the Statistical Analysis and Computing Branch, Environmental Statistics and Information Division, Office of Policy, Planning, and Evaluation, United States Environmental Protection Agency, Washington, DC under a Cooperative Agreement Number CR-825506. The contents have not been subjected to Agency review and therefore do not necessarily reflect the views of the Agency and no official endorsement should be inferred.

1. Introduction

Remote sensing has been a vastly under-utilized resource involving large amounts of investments at the national level. Even when utilized, its credibility has been at stake, largely because of the lack of tools that can assess, visualize, and communicate accurately and reliably in real-time and at the desired confidence levels.

Consider an imminent 21st century scenario: What message does a remote sensing-derived land cover land use map have about the large landscape it represents? And at what scale? ... Does the spatial pattern of the map reveal any societal, ecological or environmental condition of the landscape? ... And therefore can it be an indicator of change? ... How do you automate the assessment of the spatial structure and behaviour of change to discover critical areas and hot spots? Is the map accurate? ... How accurate is it? How do you assess the accuracy of the map? Of the change map over time for change detection? What are the implications of the kind and amount of change and accuracy on aspects that matter, whether it be climate, carbon emission or water resources? And with what confidence and credibility? Answers to these kinds of questions that involve multicategorical raster maps based on remote sensing and other geospatial data are now overdue.

2. Classified Geospatial Raster (Cell by Cell) Maps and Setting the Geospatial Raster Stage

Geospatial data form the foundation of an information-based society. An urgent need for today is to achieve credible raster maps of societal, ecological and environmental variables to facilitate quantitative characterization and comparative analysis of subregions of concern. The methodological toolbox and the software toolkit should support core missions of government agencies as well as their interactive counterparts in society. The single and multiple raster map contexts include formulation and evaluation of policy at national, state and local levels, crisis management and protection of the societal infrastructure.

The information-based global economy is becoming a geospatial information-based economy. Such tools as aerial and satellite remote sensing imagery, the Global Positioning System (GPS), and

Geographic Information Systems (GIS) are revolutionizing the conduct of business, science and government alike. Geospatial information is increasingly becoming the driving force for decision making across the local to the global continuum. Tasks as varied as planning urban growth, managing a forest, implementing 'precision farming', assessing insurance claims, setting up an automatic teller machine, drilling a well, assessing groundwater contamination, designing a cellular phone network, guiding 'intelligent' vehicles, assessing the market for manufactured goods, managing a city, operating a utility, improving wildlife habitat, monitoring air quality, assessing environmental impact, designing a road, studying human health statistics, minimising water pollution, undertaking real-estate transactions, preserving wetlands, mapping natural hazards and disasters, providing famine relief or studying the causes and consequences of global climate change can be enhanced greatly by the use of some form of geospatial technology and the resultant map products.

Cell-based raster data sets or grids are especially well suited to represent traditional geographic phenomena such as elevations, slope, precipitation that vary continuously. Raster data sets can also be used to represent the less traditional types of information such as population density, consumer behaviour and other demographic characteristics. In addition, grids are the ideal method for data representation for spatial modelling applications of land cover, land use characteristics and relevant attributes such as hydrologic modelling or evaluating the dynamics of population change over time. While the classified raster map analysis is largely raster-based, vector data and vector polygon representation and research also finds its natural place in the geospatial mapping and analysis effort.

Interagency coordination is under way almost everywhere to produce a series of nationally consistent data products and to define a series of standards that are consistent with the key uses of remotely-sensed data. Compliance with these standards will allow a variety of initiatives to collect and process remote sensing data to produce a variety of products that can be interchanged, linked and compared across regions. Developments in science and technology are expected to provide new opportunities for collecting and organising data that will expand our capabilities of integrating data and programmes across resources, agencies and temporal and spatial scales greatly.

3. Ecosystem Health Assessment with Remote Sensing Data

Ecosystem health entails both status and trends. Ecosystem processes operate in space and time, so not only what is observed and how it is observed are important, but also where and when the observations are made. Ecosystems are open systems with various processes functioning in gradients over a range of spatial and temporal scales. Ecological hierarchy theory implies that observations made in a particular spatial and temporal frame of reference will find certain processes more strongly expressed than others.

Ecosystem health has been evolving over the past decade from explorations of the possible relevance of the concept of health at the ecosystem and landscape scale (Nielsen 1999) to the exploration of the concept regarding the monitoring and assessment of large-scale ecosystems (Rapport *et al.* 1995) to an ever-widening range of considerations. These include incorporating societal values (Rapport *et al.* 1998) at ecosystem and landscape scales, economic and social determinants and the consequences of ecological conditions (Buckingham 1998; Rapport *et al.* 1998) and the increasing dependence of human health on ecosystem conditions (Epstein 1995; Epstein and Rapport 1996; Huq and Colwell 1996; Karr 1997; McMichael 1997). It also includes the encouragement of appropriate techniques for rapid, accurate and economically feasible methods for regional scale assessment of ecological conditions including developments in GIS and remote sensing (Patil and Myers, 1999; Rapport 1999).

We focus on the new generation of techniques and analyses that permit broad scale geographical assessments of environmental and ecological conditions. These techniques, including recent advances in the applications of GIS and remote sensing imagery, will prove invaluable for advances in ecosystem health. These tools may be the basis for assessing ecosystem health parameters (for instance, biotic community structure, biotic cover and primary productivity) in relation to the provision of ecosystem services at regional scales (for example, nutrient flux in watersheds, sediment loads to drainage basins, biodiversity expressed in quantities and ranges of habitat). These techniques will drive the next generation of quantitative assessment of ecosystem health at regional scales (Rapport 1999).

While considerable effort has been devoted to the identification of stressors and to the development of indicators, protocols for framing observations spatially and temporally are less well established.

No one has yet been heroic enough to attempt the delineation of ecosystems, but there are several competing strategies for ecological mapping that can help to segregate areas with respect to the degree of coupling among biophysical processes. All such mapping strategies lead to hierarchies of zones nested over several levels of spatial detail. The successful application of such ecological mapping strategies requires information on the distribution of natural and anthropogenic features across landscapes. Making *in situ* determinations everywhere is not feasible, so remote sensing constitutes the only available method by which to establish geographic context and achieve spatial integration. Landscape ecology provides increasingly mature guidance regarding spatial organisation among ecosystems and how remotely sensed data can be used to understand such organisation in a particular geographic setting.

Resolution, frequency of coverage, cost and procurement are not the only concerns with respect to the use of remotely sensed data for the assessment of ecosystem health. Remotely sensed data seldom constitute a complete information source for ecosystem analysis. Improvements in these respects will not necessarily translate into substantially improved monitoring and assessment unless we learn better ways of incorporating pixelized spectral data into a multi-tiered analysis that integrates intensive site studies, distributed sample plots and various partial coverages from remote sensors at different resolutions and different times. In the remote-sensing community, this is known as the challenge of data fusion. The statistical, environmental, and ecological communities know the problem by a variety of other names. Remotely sensed data is also problematic in the sense that the variates (spectral bands) are not measures that speak to a particular focus of interest like the variables conventionally measured in field surveys. They literally constitute a picture of the landscape and many different sorts of things can be seen from the same picture. Extraneous information is noise to a particular analysis; segregation of relevant variability from irrelevant variability often requires sophisticated methodology that is the special domain of image analysis. Contemporary improvements in sensors tend to compound such difficulties. A related concern is for parsimony. Since a massive infusion of high-resolution hyper-spectral image data could swamp most environmental data analysis facilities, it becomes important to exploit opportunities for compression. Furthermore, the remote sensing community is only beginning to address spatial analysis explicitly in its routine work with image data. Ironically, the spatial structure of image data may ultimately be of the greatest importance for ecosystem analysis.

While compilations of land use and land cover maps by remote sensing have a long history, this basic process is seldom easy or even satisfying with respect to desired detail and accuracy of classification. Determining the occurrence and character of landscape change has proven equally problematic. The specific issues are deep but concrete and finite.

4. Landscape Patterns and their Comparison

When a landscape is represented by multiple land cover types, instead of just forest/non-forest, the challenge is to define a measurement of landscape fragmentation that can be applied to any defined geographic area. Such a measurement would ideally allow quantitative decision-making for determining when a landscape pattern has changed significantly, either within the same geographic extent over time, or amongst different locations within a similar ecoregion. Of importance to this exercise can be the ability to identify ecosystems such as may be delineated by watershed boundaries that are close to the critical point of transition (Figure 7.1) into a different, possibly degraded, ecosystem where the landscape matrix has become developed land (Figure 7.1d), supporting only small, sparsely scattered forest islands that do not provide sufficient forest interior habitat. Along with the collapse of forest-interior species richness, degradation may also be attested to by the increasing environmental contamination that is also associated with intensive land development.

Such a measurement of landscape fragmentation can then be a primary component of an ecosystem risk assessment in a manner addressed by Graham *et al.* (1991). Indentifying areas whose landscape level ecosystems are poised for a great reduction in overall species and/or the elimination of critical functional groups is of utmost concern because such areas may still be salvaged with intervention by proper land use planning. Meanwhile, other areas that have 'crossed the line' but are not too beyond the critical point may still be reversible. Indeed, ecosystems that are near critical transition points present both risks and opportunities.

As a Markov transition model, the landscape fragmentation generating model can be described fully by its transition probability

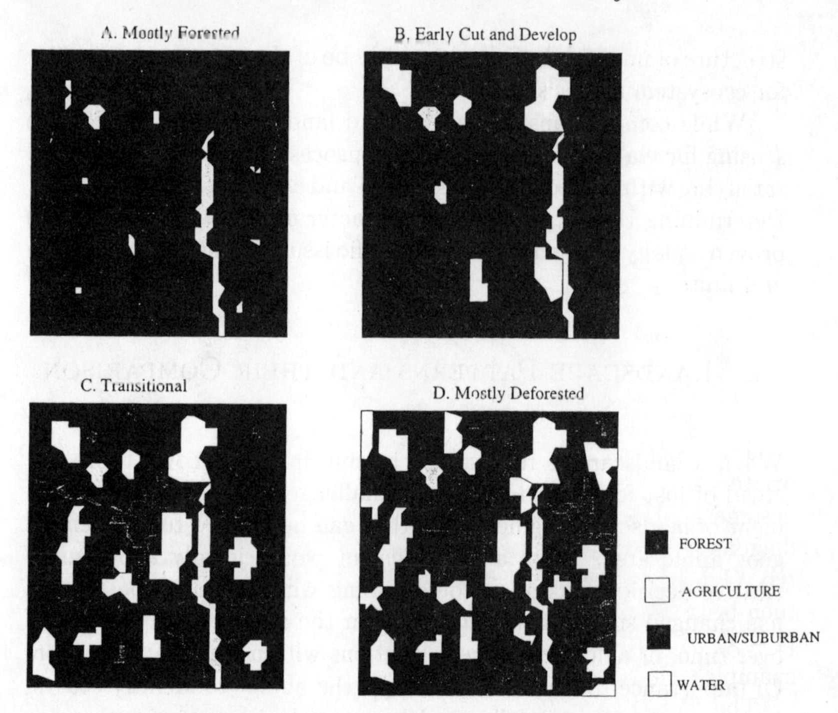

Figure 7.1: Schematics of Landscape Fragmentation

matrices. To simulate the null scenario of a self-similar fragmentation process at each resolution, we may invoke a stationary model whereby the same stochastic matrix applies at each transition. Stationary probability transition matrices are based on the characteristics of actual watershed-delineated landscapes. These landscapes represented by 8 land cover types at a floor resolution of 30 meter pixels.

The actual landscape maps are reproduced in Figure 7.2. More details about the data sources can be found through the Web page (*www.pasda.psu.edu*), which includes metadata for the land coverage. The Sinnemahoning Creek watershed is mostly forested, representing a continuum of forest interior wildlife habitat. The Jordan Creek watershed represents a transitional landscape that barely maintains a connected forest matrix which is encroached upon by agriculture and urban/suburban land use. Meanwhile, the Conestoga Creek watershed represents a landscape that is dominated by open agricultural land and highly aggregated urban/suburban land, with isolated patches of remaining forest. Conditional entropy profiles, measuring landscape fragmentation, appear in Figure 7.3.

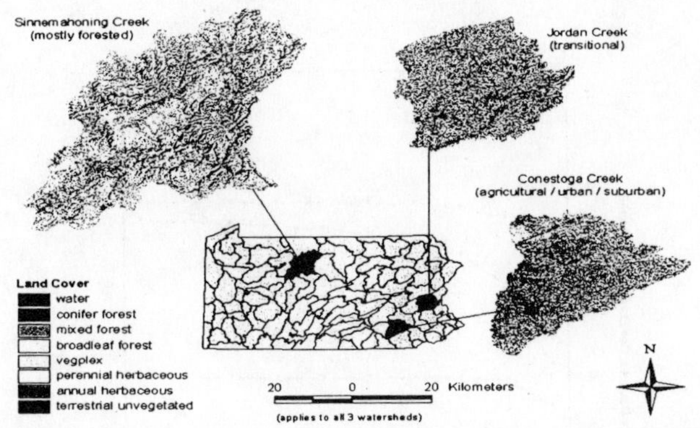

Figure 7.2: Land cover Maps for Three Watersheds of Pennsylvania

The stochastic transition matrices can be modelled as appropriate. For example, null landscape models may be obtained by designating a degree of within-patch coherence by the magnitude of diagonal elements (self-preserving probabilities) in a stochastic matrix. Labelling the diagonal value as λ, off-diagonal elements may then be distributed evenly amongst the remaining probability mass $(1 - \lambda)$ within each row. The conditional entropy profiles for some examples of such models are presented in Figure 7.4.

The shape of a conditional entropy profile appears to be governed largely by two aspects of land cover pattern, as seen in the poor resolution data: the marginal distribution, viewed as the relative frequency of each land cover; and the spatial distribution of land cover types across the given landscape. For more information, see Johnson and Patil (1998), Johnson *et al.* (1999, 2001b), Myers *et al.* (1997) and Patil (1998 a, b).

5. Classified Raster Map Modelling and Simulation with Hierarchical Markov Transition Matrix Models

The proposed approach uses a series of Markov transition matrices to generate a hierarchy of categorical raster maps at successively finer resolutions. Each transition may use a different matrix, thereby modelling distinct as well as smoothly ranging scaling domains. Even when data is available at only the finest resolution, the model is nonetheless identifiable; parameters can be estimated by exploiting a duality between hierarchical transitions in the model and spatial transitions at varying distance scales in the data map. See Johnson

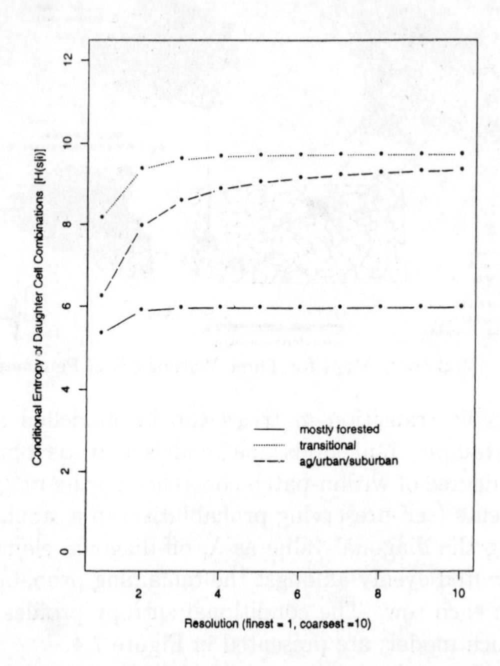

Figure 7.3: Conditional Entropy Process Profiles as Landscape Fragmentation Profiles for HMTM Models whose Transition Matrices are Obtained from Watersheds with three Distinctly Different Land cover Patterns

(1999), Johnson and Patil (1998), Johnson *et al.* (1998, 2001a, b), Patil *et al.* (2000) and Patil and Taillie (1999, 2000a).

5.1 Spatial Dependence, Auto-Association and Adjacency Matrix

While very different, our approach to the modelling of classified maps is conceptually similar to the variogram/covariogram characterization of spatial dependence employed in geostatistics (Cressie, 1991; Myers, 1982). Our goal is developing methodology for the analysis of multi-categorical map data which has the computational ease and convenience of geostatistics for numerical spatial data. We want the underlying model of spatial dependence to be a true probability model instead of the moment model of kriging.

Consider a raster map of an attribute A having κ categorical levels denoted by $a_1, a_2, \ldots a_\kappa$. For an empirical description of the spatial dependence at varying distances in the map, we employ a series $\hat{R}_0, \hat{R}_1, \hat{R}_2, \ldots$ of $\kappa \times \kappa$ matrices. The matrix $\hat{R}_n$ is obtained by

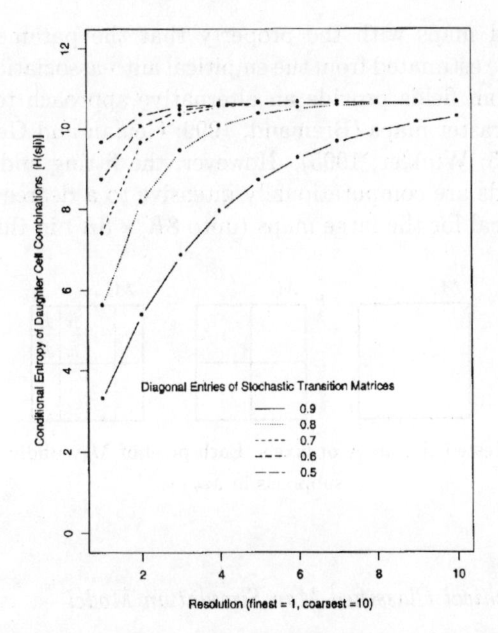

Figure 7.4: Conditional Entropy Process Profiles for HMTM Models whose Hypothetical $k \times k$ Transition Matrices have the Value λ along the Diagonal and the Value $(1-\lambda)/\ (k-1)$ off the Diagonal. Here $k=8$ and the Values of λ are Indicated in the Legend. The Stationary Vector is Uniform Across the k Categories and is used as the Initial Vector in the Model. The Floor Resolution Maps $\mathcal{G}_L$ were $1,024 \times 1,024$ (i.e., $L=10$). Large Values of λ Result in Strong Spatial Dependence (Indicated by large Profile Relief) which Persists at Larger Distances (Indicated by a Slowly Rising Profile). All Models have the same Stationary Vector and therefore the same Horizontal Asymptote

scanning the map and examining pairs of pixels which are 2^n pixels apart, either horizontally or vertically. The pixels are adjacent when $n=0$; they have 2^n-1 pixels between them for general n. The i, j entry of $\hat{R}_n$ is the relative frequency of occurrence of response (a_i, a_j) in such pairs of pixels. By definition, $\hat{R}_n$ is a symmetric matrix and its κ^2 entries sum to unity. Thus, $\hat{R}_n$ is a probability table expressing empirically the auto-association of attribute A at distance 2^n across the map. The series, $\hat{R}_0, \hat{R}_1, \hat{R}_2, \ldots$ of auto-association tables is a categorical counterpart of the empirical variogram for numerical response data.

Our next step is to develop a parametrized probability model for classified maps with the property that the parameters of the model can be estimated from the empirical auto-association matrices. Gibbs random fields provide an alternative approach to modelling categorical raster maps (Bremaud, 1999; Geman and Geman, 1984; Guyon, 1995; Winkler, 1995). However, the fitting and simulation of these fields are computationally intensive to a degree that would be impractical for the large maps (upto $8K \times 8K$) in this proposal.

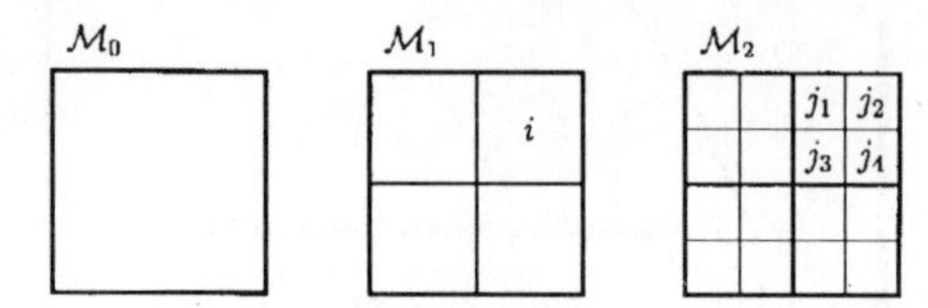

Figure 7.5: Nested Hierarchy of pixels. Each pixel of M_n subdivides into four subpixels in M_{n+1}.

5.2 *Hierarchical Classified Map Simulation Model*

The Hierarchical Markov Transition Matrix (HMTM) model generates a sequence $\mathcal{M}_0, \mathcal{M}_1, \ldots \mathcal{M}_L$ of categorical raster maps. Each map covers the same spatial extent; but successive maps are of increasingly finer resolution. The first map $\mathcal{M}_0$ consists of a single pixel and, recursively, the pixels of $\mathcal{M}_n$ are bisected horizontally and vertically to produce the pixels of $\mathcal{M}_{n+1}$, giving rise to a 'quadtree' (Samet 1990, see Figure 7.5). In describing the transition from $\mathcal{M}_n$ to $\mathcal{M}_{n+1}$, we refer to a pixel in $\mathcal{M}_n$ as a 'mother' pixel and its four subpixels in $\mathcal{M}_{n+1}$ as 'daughter' pixels.

Mapping categories are assigned to pixels of $\mathcal{M}_n$ using Markov transition matrices. We consider k mapping categories (values), labelled as 1, 2, ..., k. At the coarsest scale, assigning a value to the single pixel of $\mathcal{M}_0$ is determined by an initial stochastic (row) vector $\mathbf{p}^{[0]}$. Having assigned values to pixels of $\mathcal{M}_n$, the assignment to $\mathcal{M}_{n+1}$ is generated by a row stochastic transition matrix,

$$\mathbf{G}^{[n,n+1]} = \left[G_{ij}^{[n,n+1]}\right], \; i,j = 1,\ldots,k$$

Consider a particular model pixel of $\mathcal{M}_n$ and let its value be i. The values j of the four daughter pixels are generated by four independent draws from the distribution specified by the ith row of $\mathbf{G}^{[n,n+1]}$. The marginal distribution of mapping categories across

$\mathcal{M}_{n+1}$ is obtained from the initial vector $\mathbf{p}^{[0]}$ via the recurrence relation,

$$\mathbf{p}^{[n+1]} = \mathbf{p}^{[n]}\mathbf{G}^{[n,n+1]}.$$

Only the final, floor resolution map $\mathcal{M}_L$ may be available for analysis. From this single-resolution map, we estimate model parameters by relating spatial scaling levels across $\mathcal{M}_L$ to hierarchical levels in the model. With suitable restrictions on the model parameters, an identifiability theorem asserts that distinct sets of model parameters correspond to distinct probability distributions on $\mathcal{M}_L$. This correspondence is accomplished by relating the eigen-decomposition of the hierarchical transition matrices to the eigen-decomposition of the spatial auto-association matrices. Model fitting is accomplished by scanning the floor resolution map to estimate auto-association matrices (Patil and Taillie 1990, 2000a).

5.3 Landscape Characterization and Discrimination

The eigen-decomposition of the transition matrix may be studied for landscape characterization and discrimination. In analogy with Principal Components, the eigen values and eigen vectors can be effective discriminators of landscape spatial pattern. For example, the (left) eigenvector corresponding to the largest eigen value is the marginal land cover distribution which accounts for much of the between-watershed variability in Pennsylvania but does not capture the within-watershed spatial pattern. The second and later eigen vectors reflect spatial pattern and (in a sense) are orthogonal to the first eigen vector. Also, according to bi-orthogonality, the second and later eigen vectors are contrasts across the different mapping categories, that is, their components sum to zero. The patterns of signs in these contrasts may be indicative of scientifically meaningful associations among mapping categories with reference to landscape fragmentation. They may also suggest appropriate groupings of the categories for simplification of thematic mappings. The sensitivity of the HMTM eigen-decomposition to classification error in the map is also of interest.

5.4 Fragmentation Profiles

The fragmentation profile is a graphic display of the persistence of spatial pattern across spatial scales. Starting from a data map, a random filter is applied iteratively to produce a sequence of generalized maps, $\mathcal{G}_0, \mathcal{G}_1, \mathcal{G}_2, \ldots, \mathcal{G}_n \ldots$, where $\mathcal{G}_0$ is the data map and $\mathcal{G}_{n+1}$ is obtained from $\mathcal{G}_n$ by application of the random filter. Specifically, each pixel x of $\mathcal{G}_{n+1}$ is the union of four pixels, in a 2×2 arrangement, from $\mathcal{G}_n$; one of the four subpixels of x is selected at random and its colour is assigned to x. Accordingly, $\mathcal{G}_1, \mathcal{G}_2, \ldots$ are stochastic maps.

Let i be the colour of a given pixel in $\mathcal{G}_{n+1}$ and (j_1, j_2, j_3, j_4) the colours of its four subpixels in $\mathcal{G}_n$. Scanning the pixels of $\mathcal{G}_{n+1}$ generates a frequency table whose cells are indexed by $(i, (j_1, j_2, j_3, j_4))$. The table has k^5 cells, some of which may be empty; but the cell frequencies are random variables due to the randomness of the filter. This randomness is removed by working with the table of *expected* frequencies which is denoted by

$$M_{n+1} \times D_n, \tag{7.1}$$

where the factor M_{n+1} refers to pixels of $\mathcal{G}_{n+1}$ and is indexed by i while the factor D_n refers to 4-tuples of pixels in $\mathcal{G}_n$ and is indexed by (j_1, j_2, j_3, j_4). Using the ANOVA decomposition for Shannon entropy, the entropy $H(\cdot)$ of the joint table (7.1) can be written as

$$\begin{aligned} H(D_n|M_{n+1}) + H(M_{n+1}) &= H(M_{n+1} \times D_n) \\ &= H(M_{n+1}|D_n) + H(D_n). \end{aligned} \tag{7.2}$$

The *conditional entropy profile* is defined to be the plot of

$$H_n = H(D_n|M_{n+1}) \text{ versus } n.$$

The conditional entropy H_n summarizes quantitatively how pixels from $\mathcal{G}_{n+1}$ fragment into subpixels in the finer resolution map $\mathcal{G}_n$. Computing the entropy of expected frequencies rather than expected entropy avoids the bias associated with the expected entropy (Basharin, 1959). The profile is an increasing function of the scale parameter n and approaches a horizontal asymptote whose value depends only on the marginal land cover distribution (See Figure 7.3). See also Johnson (1999), Johnson and Patil (1998), Johnson *et al.* (2001a, b, 1999, 1998), Patil *et al.* (2000), and Patil and Taillie (1999)].

The decomposition (7.2) leads to an algorithm for computing the conditional entropy H_n without first obtaining the generalized maps

$\mathcal{G}_n$ or the joint table (7.1):

$$H_n = H(D_n) + H(M_{n+1}|D_n) - H(M_{n+1}). \tag{7.3}$$

The last term equals the entropy of the marginal land cover distribution and does not change with n, the middle term is computable from the random filter, and the expected 4-tuple frequency table D_n can be obtained recursively from the data map.

These profiles are multiscale expressions of the fragmentation pattern in the map. Their capability may be examined for the purpose of characterizing and discriminating watersheds in the region of interest. In addition, profile sensitivity to classification error in the land cover map will be of interest. It may also be interesting to study profile responsiveness to the variation of parameter values in the HMTM model. Variation in the eigenstructure of the HMTM matrices can be of particular interest.

5.5 Simulation Modelling

Maps can be generated quite rapidly using the HMTM model providing an excellent vehicle for model-based inference in categorical map analysis. Three classes of questions arise:

5.5.1. Monte Carlo Determination of the Null Distributions for Hypothesis Testing.
This includes goodness-of-fit tests and nested tests for parameter reduction as well as tests of scientific hypothesis such as self-similarity and distinct scaling domains. Due to spatial dependence, the null distributions are expected to be heavily dependent on the model parameters and need to be determined on a case-by-case basis, using estimated parameters or noisy versions thereof.

5.5.2. Determining the Distribution of Parameter Estimates and, More Importantly, of Proposed Landscape Metrics.
Assessment of meta-population variability is important for comparing metrics computed empirically on different landscapes.

5.5.3. Assessing the Responsiveness of Proposed Landscape Metrics to Differences in Landscape Structure.
By systematically changing model parameters at the same or different stages in the hierarchy, landscape structure can be varied in a controlled fashion and corresponding metric changes computed. In fact, the original purpose in developing the HMTM model was to study the responsiveness of the landscape fragmentation profile.

Calculation on actual landscapes has suggested strong correlations among many of the metrics (Hargis *et al.* 1998, Johnson 1999, Ritters *et al.* 1995). Such correlations and redundancies can be examined more effectively in a controlled simulation study rather than through an observational study.

6. Classified Raster Map Analysis for Assessment of Accuracy and Change Detection of Land Cover and Land use Maps

The need for assessing the accuracy of land cover and land use maps has become recognized universally. With the increasingly widespread application of GIS, such assessments become even more pressing. In addition to the research on one-point-in-time maps, the theoretical and methodological developments need attention to change-detection maps. Change detection is a valuable and extensively used remotely sensed data tool spanning global, national, state and local scales. Remote sensing provides temporally frequent and spatially complete coverage which may be exploited for early detection of environmental problems, insect or disease outbreaks, epidemiologic problem conditions, fire or fire risks, etc.

A model for accuracy assessment may be formalized as follows. Each pixel in the map carries two categorical values (for example, land cover types) denoted by d and t, where t is the 'true' or 'reference' value and d is the data value assigned by the classification algorithm. A population error matrix results from scanning the map and recording the proportion π_{dt} of pixels that carry the vaules (d, t). Here, d and t range from 1 to k, where k is the number of categories. In most cases, the reference value t is available for only a sample of pixels and an estimate $\widehat{\pi}_{dt}$ is obtained from this sample. The error matrix (Figure 7.6) serves as the basis for description as well as further analyses of accuracy (Congalton 1991, Congalton and Green 1999, Khorram *et al.* 1999, Lunetta and Elvidge 1998, Patil and Tallie 2000b, c). Inferential interest lies in the entire matrix $\{\pi_{dt}\}$ and also in several parametric scalars and vectors such as $\kappa = (\mathcal{P}\pi_{ii} - \mathcal{P}\pi_{i.}\pi_{.i})/(1 - \mathcal{P}\pi_{i.}\pi_{.i})$, the Kappa coefficient of agreement; $\{\pi_{tt}/\pi_{.t}\}$, producer's accuracy; $\{\pi_{dd}|\pi_{d\cdot}\}$, user's accuracy and $\{\pi_{.t}\}$, proportion of areal extent in each class.

A cost-effective and efficient accuracy assessment protocol requires innovations in sampling design and analysis. Improved theory and methods are needed for single map accuracy assessment as well

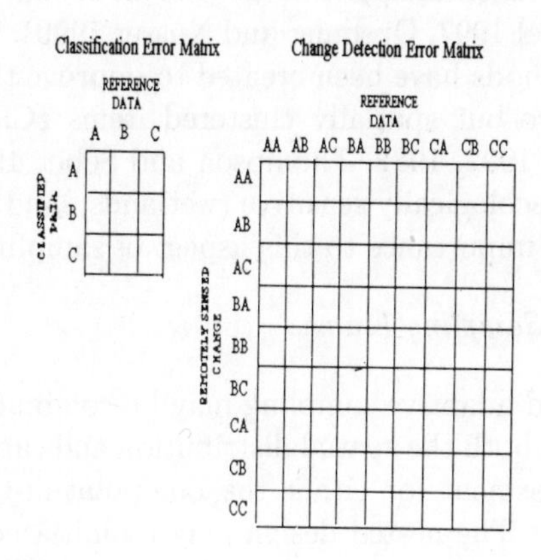

Figure 7.6: A Comparison between a Single Classification Error Matrix and a Change Detection Error Matrix for the same Vegetation/Land use Categories

as accuracy assessment of change detection. The sampling design must permit precise estimation for rare classes (for example, a land cover class or a particular type of change) and be geographically well distributed to improve precision of overall and subregional estimates. The performance of different sampling designs under different spatial patterns of error needs evaluation. The major research themes on analysis may include change detection error matrix analysis, multiscale bivariate map analysis of change and model-based accuracy assessment.

6.1 Sampling Design

Authors Stehman (2000, 1999a, b, c,, 1997a, b, 1996a, b) and Congalton and Green (1999) have reviewed fundamental sampling designs commonly used in current practice. Sampling designs applicable when there is some prior knowledge of the spatial location of rarity include the stratified Neyman optimum allocation with fine-tuning variants such as disproportionate sampling (Biging *et al.* 1998, Chrisman 2003, 2000, Kalton and Anderson 1986) and special effort sampling (Khorram *et al.* 1999). The core of the sampling design research may focus on two relatively recent design innovations: Markov chain sampling and adaptive cluster sampling. Markov chain sampling has been developed to enhance the spatial properties of samples in environmental applications (Breidt 1995a, b, Fuller 1999, Nusser and Goebel 1997, Opsomer and Nusser 1999). Adaptive cluster sampling methods have been created to improve the efficiency of estimates for rare but spatially clustered items (Christman 2003, 2000, Thompson 1992, 1982, Thompson and Seber 1996). The rare classes are often ecologically sensitive (wetlands, land cover change), lending strategic importance to this aspect of sampling design.

6.2 Nested Area Sampling Frames

Markov chain and adaptive sampling may be combined via a nested design to address both the spatial distribution and rare class features of accuracy assessment for either the one-point-in-time or change detection setting. The nested design may be initiated by a Markov chain. Suitable variants may be examined, consistent with the stipulated error rates and spatial patterns representative of ecosystems of varying complexity: agriculture, rangeland, forest, etc. (Congalton 1988). The performance of the sampling designs and the related data analysis techniques may be investigated on real and simulated data sets provided for classified maps and reference databases. Comparative performance studies also need to be conducted relative to the disproportionate sampling/special effort sampling protocols involved. Refer Figure 7.7.

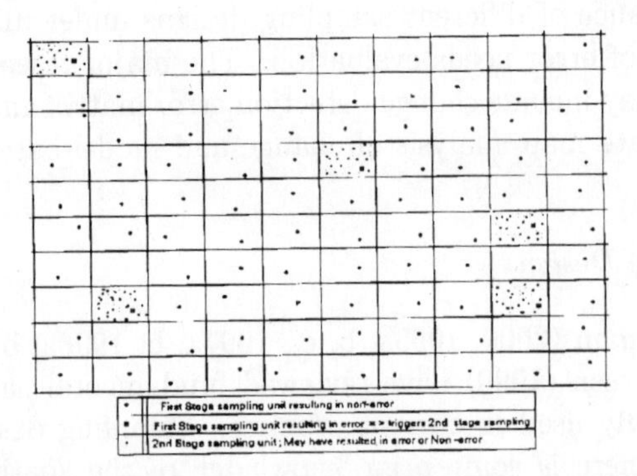

Figure 7.7: Simulation Result showing a Possible Two-Stage Adaptive Sample

6.3 Spatial Patterns of Classification Errors in Thematic Maps and their Applications to Sampling Designs for Accuracy Assessment

Because the spatial distribution of classification error affects performance of the sampling designs, further analysis is needed to determine which options are more efficient certain scenarios of a spatial pattern. The effect of spatial pattern on the estimation of the error matrix and associated parameters is particularly important. Two questions arise:

(a) How does spatial pattern affect estimator precision (performance) of a given sampling design?

(b) How can one exploit a suspected pattern to arrive at a design that achieves high performance? Refer Figure 7.8 (Congalton 1988).

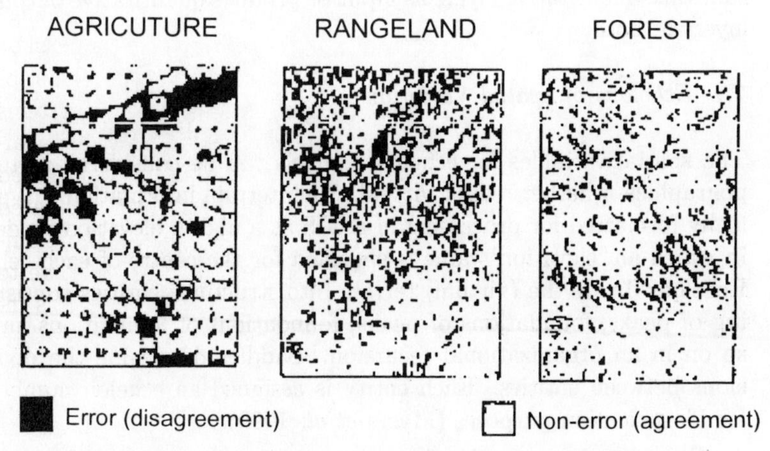

Figure 7.8: Spatial Patterns of Error for Three Ecosystems

7. Analyzing Spatial Variation in Quantitative Data and Determining Contexts of Temporal Changes and Class Errors using Echelon Analysis

Quantitative spatial data are important inputs of many environmental process models for determining the future implications of current resource use, policies and interventions. The end-products of applying such models are often mappings of indexes for the level of potential environmental impact, which then become guides to the allocation of economic and technical resources for ameliorating environmental damage. Errors in quantitative spatial data layers will propagate through environmental models and find expression in the resulting indexes of environmental impact. However, the consequences of such errors for decision-making may well depend upon the spatial pattern and the location of the errors.

It is therefore desirable to have a systematic means of determining spatial organization in mappings of quantitative variables. Echelons present a means for determining objectively quantitative spatial structure for direct mapping either with or without computer-assisted visualization (Johnson *et al.* 1998, Kurihara *et al.* 2000, Myers *et al.* 1999, 1997, 1995, Patil and Taillie 1999, Ramakomud 1998, Smits and Myers 2000). Thus, they can facilitate the analysis of the implications of errors associated with environmental models that take quantitative layers as input or produce quantitative output layers or both.

7.1 Echelons of Spatial Variation

The spatial variables for echelon analysis can be considered as topographies, whether real or virtual. Such terrain information is typically formatted for processing in a GIS as a digital elevation model in which an 'elevation' value is specified for the centre of each cell. Echelons divide the (virtual) terrain into structural entities consisting of peaks, foundations of peaks, foundations of foundations and so on in an organizational recursion. Saddles determine the divisions between entities. Each entity is assigned an echelon number for identification purposes (Myers *et al.* 1999).

Consider, for example, the terrain depicted in profile with division as seen in Figure 7.9a. The numbered entities thus determined are called echelons. Echelons are determined directly by organizational complexity in the spatial variable and not by either absolute 'elevation' or steepness. Echelons form an extended family of terrain entities and determine a family tree as illustrated in Figure 7.9b. This is a 'scaled tree' in the sense that the height of each vertical edge corresponds to the height of the echelon above its founder. The acummulated height above the root is the height of the terrain. The number of 'ancestors' for an echelon is a local measure of regional complexity. The echelons also comprise a structural hierarchy of organizational orders. These orders are assigned and numbered in the same way as for a network of streams and tributaries (Rodriguez-Iturbe and Rinaldo, 1997).

A suite of form attributes can be determined for each echelon, including the area extent of the basal slice and vertical projection

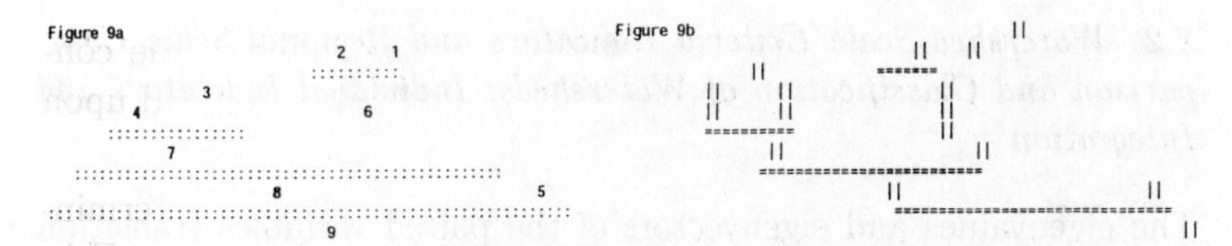

Figure 7.9: Echelons of Spatial Variation

above its founder. Some form attributes may depend upon an interval scale of measure for the vertical dimension, but the echelon decomposition requires only an ordinal scale of measurement. A standard table of echelon characteristics contains a record with 10 fields for each echelon, including echelon ID number, order, founder, maximum level, minimum level, relief, cells, progeny, ancestors, and setting within the tree. The table is associated with an echelon map file giving the 'level' value and echelon ID number for each cell. Echelons thus formalize the structural complexity of the surface variable without incurring any loss of information with respect to the surface level. Since most echelon trees are much too complicated for visual study as dendrograms, characterization and comparison of echelon trees is done through analytical processes such as pruning (Myers *et al.* 1999).

8. Integrated Regional Assessment, Model Prediction and Regional Scale Comparison Involving Classified Raster Maps

8.1 *Effects of Classification Error on Model Predictions*

A few recent studies have attempted to study the effects of classification and its attendant errors on model predictions involving thematic maps derived from remote sensing data. Outputs studied include biomass, landscape pattern metrics, leaf area index, local climate interpolation for agriculture, net carbon release, net primary production, non-point source water pollution, species diversity map, etc. See, for example, Finn (2000), Foody *et al.* (1996), Kyriakidis and Dungan (2001), Lo and Faber (1997), Moisen and Edwards (1999), Riley *et al.* (1997) and Wickham *et al.* (1997). A systematic study of the model sensitivity issues pertaining to the thematic map and assessment of its accuracy is needed. Initial emphasis is expected to be placed on the model outputs which are landscape pattern metrics, thematic maps and minimum-mapping-unit-based characteristics.

8.2 Watershed-Scale Criteria Indicators and Regional-Scale Comparison and Classification of Watersheds: Individual Indicators and Integration

The eigenvalues and eigenvectors of the paired variables transition matrices provide a suite of watershed-specific indicators as in the case of single maps. In order to characterize, compare and classify watersheds of a region according to their health, distress and degradation and vulnerability for the purposes of conservation, restoration, intervention, etc., there is a need to develop, adapt and fine-tune available techniques to combine these indicators into a meaningful number of composite criteria indicators using broadly recognized approaches.

Besides the problem of correlations and uncertainty, the task of synthesizing indicators faces several other difficulties such as the relative importance of different indicators and the different impacts on human beings and ecosystems. There are several approaches which might be used for this task, for example, multiple criterion decision modelling (Filar *et al.* 1999, Olson *et al.* 1996 and Zeleny 1982), fuzzy set theory (Dubois and Prade 1980, Zimmerman 1987), artificial intelligence (Rauscher and Hacker 1989) and spatial information integrating technology (Osleeb and Kahn 1999).

8.3 Integration of Individual Indicators

This important integration effort can benefit from a combination of the Analytic Hierarchy Process (AHP) approach (Saaty 1999) with a method of fuzzy ranking (Tran and Duckstein 2000). It has the potential of providing an appropriate means to deal with the problems of relative contributions of different indicators as well as their uncertainty and correlations.

9. CONCLUSION

Information technologies promise to make governance more efficient and responsive. Information technologies capable of being used for credible raster map analysis and change detection are needed for an integrated regional assessment of environment and development in the 21st century.

Our approach offers new orders of synthesis by addressing quantitatively and objectively the automated spatial expression of social, ecological and environmental indicators. It offers new mathematical probes for spatial characteristics in classified thematic mappings and impact indicators. It provides for the necessary next steps in moving from the characterization of complex landscapes to comparative analysis of the status of ecosystems interfacing in space and time. And it will help alleviate the unfortunate syndrome of drowning-in-information and hungry-for-wisdom by providing a working toolbox for the integration and synthesis. Everyone concerned with multi-scale regional assessment may therefore be able to detect pathological changes on a synoptic basis as a pre-requisite for enlightened environmental policy at watershed and ecoregion scales.

REFERENCES

Basharin, G.P. (1959), 'On a Statistical Estimate for the Entropy of a Sequence of Independent Random Variables', *Theory of Probability and its Applications*, 4, pp. 333-6.

Biging, G.S., D.R. Colby and R.G. Congalton (1998), 'Sampling Systems for Change Detection Accuracy Assessment', in R.S. Lunnetta and C.C. Elvidge (eds.), *Remote Sensing Change Detection: Environmental Monitoring Methods and Applications*, Chelsea, MI: Ann Arbor Press.

Breidt, F.J. (1995a), 'Markov Chain Designs for One-per-stratum Sampling', *Survey Methodology*, 21, pp. 63-70.

Breidt, F.J. (1995b), 'Markov Chain Designs for One-per-stratum Spatial Sampling', in *Proceedings of the Section on Survey Research Methods*, Washington D.C.: American Statistical Association.

Bremaud, P. (1999), *Markov Chains: Gibbs Fields, Monte Carlo Simulation and Queues*, New York: Springer-Verlag.

Buckingham, D.E. (1998), 'Does the World Trade Organization Care About Ecosystem Health? The Case of Trade in Agricultural Products', *Ecosystem Health*, 4, pp. 92-108.

Christman, M.C. (2003), 'Adaptive Two-stage One-per-stratum Sampling, *Environmental and Ecological Statistics*, 10, (to appear).

Christman, M.C. (2000), 'A Review of Quadrat-based Sampling of Rare, Geographically Clustered Populations', *Journal of Agicultural, Biological and Environmental Statistics*, 5, pp. 168-201.

Congalton, R. (1991), 'A Review of Assessing the Accuracy of Classification of Remotely Sensed Data, *Remote Sensing of the Environment*, 37, pp. 35-45.

Congalton, R.G. (1988), 'Using Spatial Autocorrelation Analysis to Explore Errors in Maps Generated from Remotely Sensed Data', *Photogrammetric Engineering and Remote Sensing*, 54, pp. 587-92.

Congalton, R.G. and K. Green (1999), *Assessing the Accuracy of Remotely Sensed Data: Principles and Practices*, Boca Raton: Lewis Publishers.

Cressie, N.A.C. (1991), *Statistics for Spatial Data*, New York: John Wiley & Sons.

Dubois, D. and H. Prade (1980), *Fuzzy Sets and Systems: Theory and Applications*, New York: Academic Press.

Epstein, P.R. (1995), 'Emerging Diseases and Ecosystem Instabilities: New Threats to Public Health', *American Journal of Public Health*, 85, pp. 168-72.

Epstein, P.R. and D.J. Rapport (1996), 'Changing Coastal Marine Environments and Human Health', *Ecosystem Health*, 2, pp. 166-76.

Filar, J.A., N.P. Ross and M.L. Wu (1999), *Environmental Assessment Based on Multiple Indicators*, Washington D.C.: CEIS, U.S. EPA.

Finn, J.T. (2000), 'Study Proposal: Accuracy Assessment of Vegetation and Biodiversity Maps of Southern New England', Manuscript.

Foody, G.M., G. Palubinskas, R.M. Lucas, P.J. Curran and M. Honzak (1996), 'Identifying Terrestrial Carbon Sinks: Classification of Successional Stages in Regenerating Tropical Forest from Landset TM Data', *Remote Sensing of the Environment*, 55, pp. 205-16.

Fuller, W.A. (1999), 'Environmental Surveys Over Time', *Journal of Agricultural Biological and Environmental Statistics*, 4, pp. 331-45.

Geman, S. and D. Geman (1984), 'Stochastic Relexation, Gibbs Distribution, and the Bayesian Restoration of Images', *IEEE Transactions on Pattern Analysis and Machine Intelligence*, 6, pp. 721-41.

Graham, R.L., C.T. Hunsaker, R.V. O'Neill and B.L. Jackson (1991), 'Ecological Risk Assessment at the Regional Scale', *Ecological Applications*, 1, pp. 196-206.

Guyon, X. (1995), *Random Fields on a Network: Modeling, Statistics and Applications*, New York: Springer-Verlag.

Hargis, C.D., J.A. Bissonette and J.L. David (1998), 'The Behavior of Landscape Metrics Commonly Used in the Study of Habitat Fragmentation', *Landscape Ecology*, 13, pp. 167-86.

Huq, A. and R.R. Colwell (1996), 'Vibrios in the Marine and Estuarine Environment: Tracking Vibrio Cholerae', *Ecosystem Health*, 2, pp. 198-214.

Johnson, G.D. (1999), 'Landscape Pattern Analysis for Assessing Ecosystem Condition: Development of a Multi-Resolution Method and Application to Watershed Delineated Landscapes in Pennsylvania', Ph.D. Thesis, University Park, Pennsylvania: The Pennsylvania State University.

Johnson, G.D. and G.P. Patil (1998), 'Quantitative Multiresolution Characterization of Landscape Patterns for Assessing the Status of Ecosystem Health in Watershed Management Areas', *Ecosystem Health*, 4, pp. 177-89.

Johnson, G.D., W.L. Myers, G.P. Patil and C. Taillie (2001a), 'Fragmentation Profiles for Real and Simulated Landscapes', *Environmental and Ecological Statistics*, 8, pp. 5-20.

Johnson, G.D., W.L. Myers, G.P. Patil and C. Taillie (2001b), 'Characterizing Watershed Delineated Landscapes in Pennsylvania Using Conditional Entropy Profiles', *Landscape Ecology*, 16, pp. 597-610.

Johnson, G.D., W.L. Myers, G.P. Patil and C. Taillie (1999), 'Multiresolution Fragmentation Profiles for Assessing Hierarchichally Structured Landscape Patterns', *Ecological Modelling*, 116, pp. 293-301.

Johnson, G.D., W.L. Myers, G.P. Patil and D. Walrath, (1998), 'Multiscale Analysis of the Spatial Distribution of Breeding Bird Species Richness Using the Echelon Approach', in P. Bachmann, M. Kohl and R.R. Paivinen (eds.), *Assessment of Biodiversity for Improved Forest Planning*, Boston: Kluwer Academic Publishers, pp. 135-150.

Kalton, G. and D.W. Anderson (1986), 'Sampling Rare Populations', *Journal of the Royal Statistical Society, Series A*, 149, pp. 65-86.

Karr, J.R. (1997), 'Bridging the Gap Between Human and Ecological Health', *Ecosystem Health*, 3, pp. 197-9.

Khorram, S., G.S. Biging, N.R. Chrisman, D.R. Colby, R.G. Congalton, J.E. Dobson, R.L. Ferguson, M.F. Goodchild, J.R. Jensen and T.H. Mace (1999), *Accuracy assessment of remote sensing derived change detection*, ASPRT Monograph Series, Bethesda: American Society for Photogrammetry and Remote Sensing.

Kurihara, K., W.L. Myers and G.P. Patil (2000), 'The Relationship of the Population and Land cover Patterns in Tokyo Area Based on Remote Sensing Data', *Community Ecology, http://www.stat.psu.edu/gpp/newpage11.htm.*

Kyriakidis, P.C. and J.L. Dungan (2001), 'Assessing Thematic Classification Accuracy and the Impact of Inaccurate Spatial Data on Ecological Model Predictions', *Environmental and Ecological Statistics*, 8, pp. 311-30.

Lo, C.P. and B.J. Faber (1997), 'Integration of Landsat Thematic Mapper and Census Data for Quality of Life Assessment', *Remote Sensing of Environment*, 62, pp. 143-57.

Lunetta, R.S. and C.D. Elvidge (1998), *Remote Sensing Change Detection: Environmental Monitoring Methods and Applications*, Chelsea: Ann Arbor Press.

McMichael, A.J. (1997), 'Global Environmental Change and Human Health: Impact Assessment, Population Vulnerability, Research Priorities', *Ecosystem Health*, 3, pp. 200-10.

Moisen, G.G. and T.C. Edwards, Jr. (1999), 'Use of Generalized Linear Models and Digital Data in a Forest Inventory of Northern Utah', *Journal of Agricultural, Biological, and Environmental Statistics*, 4, pp. 372-90.

Myers, D.E. (1982), 'Matrix Formulation of Co-kriging', *Journal of the International Association for Mathematical Geology*, 14, pp. 249-57.

Myers, W.L., G.P. Patil and K. Joly (1997), 'Echelon Approach to Areas of Concern in Synoptic Regional Monitoring', *Environmental and Ecological Statistics*, 4, pp. 131-52.

Myers, W.L., G.P. Patil and C. Taillie (1999), 'Conceptualizing Pattern Analysis of Spectral Change Relative to Ecosystem Health', *Ecosystem Health*, 5, pp. 285-93.

Myers, W.L., G.P. Patil and C. Taillie (1995), 'Comparative Paradigms for Biodiversity Assessment', in T.J. Boyle and B. Boontawee (eds.), *Measuring and Monitoring Biodiversity in Tropical and Temperate Forests*, Indonesia: CIFOR, Bogor.

Nielsen, N.O. (1999), 'The Meaning of Health', *Ecosystem Health*, 5, pp. 65-6.

Nusser, S.M. and J.J. Goebel (1997), 'The National Resources Inventory: A Long-term Multi-resource Monitoring Programme', *Environmental and Ecological Statistics*, 4, pp. 181-204.

Olson, J.M., D.G. Brown, D.J. Campbell and L. Berry (1996), 'A Hierarchical Approach to the Integration of Social and Physical Data Sets: The Rwanda Society Environmental Project', *Human Dimensions Quarterly*, 4, pp. 14-17.

Opsomer, J.D. and S.M. Nusser (1999), 'Sample Designs for Watershed Assessment', *Journal of Agricultural, Biological and Environmental Statistics*, 4, pp. 429-42.

Osleeb, J.P. and S. Kahn (1999), 'Integration of Geographic Information', in V.H. Dale and M.R. English (eds.), *Tools to Aid Environmental Decision Making*, New York: Springer-Verlag.

Patil, G.P. (1998a), 'Environmental and Ecological Regional Policy Research with Remote Imagery and Geospatial Information, Issues, Approaches, and Examples' Technical Report 98-1201, Center for Statistical Ecology and Environmental Statistics, Department of Statistics, Pennsylvania State University, University Park.

Patil, G.P. (1998b), 'Statistical Ecology and Environmental Statistics for Cost-effective Ecological Synthesis and Environmental Analysis', in R.S. Ambasht (ed.), *Modern Trends in Ecology and Environment*, The Netherlands: Backhuys Publishers.

Patil, G.P., G.D. Johnson, W.L. Myers and C. Taillie (2000), 'Multiscale Statistical Approach to Critical-area Analysis and Modelling of Watersheds and Landscapes', in C.R. Rao and G.J. Szekely (eds.), *Statistics for the 21st Century: Methodologies for Applications of the Future*, New York: Marcel Dekker.

Patil, G.P. and W.L. Myers (1999), 'Environmental and Ecological Health Assessment of Landscapes and Wateshed with Remote Sensing Data', *Ecosystem Health*, 5, pp. 221-4.

Patil, G.P., W.L. Myers, Z. Luo, G.D. Johnson and C. Taillie (2000), 'Multiscale Assessment of Landscapes and Watersheds with Synoptic Multivariate Spatial Data in Environmental and Ecological Statistics', *Mathematical and Computer Modeling*, 32, pp. 257-72.

Patil, G.P. and C. Taillie, (2000a), 'A Multiscale Hierarchical Markov Transition Matrix Model for Generating and Analyzing Thematic Raster Maps', Technical Report 2000-0603, Center for Statistical Ecology and Environmental Statistics, Department of Statistics, Pennsylvania State University, University Park.

Patil, G.P. and C. Taillie, (2000b), 'Modeling and Interpreting the Accuracy Assessment Error Matrix for a Doubly Classified Map', Technical Report 99-0502, Center for Statistical Ecology and Environmental Statistics, Department of Statistics, Pennsylvania State University, University Park.

Patil, G.P. and C. Taillie, (2000c), 'Analytic Solution of the Regularized Latent Truth Model for Binary Maps', Technical Report 2000-0601, Center for Statistical Ecology and Environmental Statistics, Department of Statistics, Penn State University, University Park.

Patil, G.P. and C. Taillie, (1999), 'A Markov Model for Hierarchically Scaled Landscape Patterns', *Bulletin of the International Statistical Institute*, 58, pp. 89-92.

Ramakomud, A. (1998), 'Change Detection Using Hyperclustered Data: The Spatial Averging Approach', *Master of Science Thesis*, Pennsylvania State University, University Park.

Rapport, D.J. (1999), 'Gaining Respectability: Development of Quantitative Methods in Ecosystem Health', *Ecosystem Health*, 5, pp. 1-2.

Rapport, D.J., R. Cotanza and McMichael (1998), 'Assessing Ecosystem Health: Challenges at the Interface of Social, Natural and Health Sciences', *Trends in Research in Evolution and Ecology*, 13, pp. 397-402.

Rapport, D.J., C.L. Gaudet and P. Calow (1995), *Evaluating and Monitoring the Health of Large-Scale Ecosystem*, Berlin: Springer-Verlag.

Rauscher, H.M. and Hacker, R. (1989), 'Overview of Artificial Intelligence Applications in Natural Resource Management', *Journal of Knowledge Engineering*, 2, pp. 30-42.

Ritters, K.H., R.V. O'Neill, C.T. Hunsaker, J.D. Wickham, D.H. Yankee, S.P. Timmins, K.B. Jones and B.L. Jackson (1995), 'A Factor Analysis of Landscape Pattern and Structure Metrics', *Landscape Ecology*, 10, pp. 23-9.

Riley, R.H., D.L. Phillips, M.J. Schuft and M.C. Garcia (1997), 'Resolution and Error in Measuring Land cover Change: Effects on Estimating Net Carbon Release from Mexican Terrestrial Ecosystems', *International Journal of Remote Sensing*, 18, pp. 121-37.

Rodriguez-Iturbe, I. and A. Rinaldo (1997), *Fractal River Basins: Chance and Self-Organization*, Cambridge: Cambridge University Press.

Rhyne, T.M. (2000), 'Scientific Visualization in the Next Millennium', *IEEE Computer Graphics and Applications*, 20, pp. 20-1.

Saaty, T.L. (1999), *Decision Making for Leaders: The Analytic Hierarchy Process for Decisions in a Complex World* (1999/2000 edition), 3rd revised edition, vol. 2, Pittsburgh: RWS Publications.

Samet, H. (1990), *Applications of Spatial Data Structures: Computer Graphics, Image Processing, and GIS*, Reading : Addison-Wesley.

Smits, P.C. and W.L. Myers (2000), 'Echelon Approach to Characterize and Understand Spatial Structures of Change in Multi-temporal Remote-sensing Imagery', *IEEE Transactions of Geoscience and Remote Sensing*, 38, pp. 2,299-309.

Stehman, S.V. (2000), 'Practical Implications of Design-based Sampling Inference for Thematic Map Accuracy Assessment', *Remote Sensing of the Environment*, 72, pp. 35-45.

Stehman, S.V. (1999a), 'Basic Probability Sampling Designs for Thematic Map Accuracy Assessment', *International Journal of Remote Sensing*, 20, pp. 2,423-41.

Stehman, S.V. (1999b), 'Comparing Thematic Maps Based on Map Value', *International Journal of Remote Sensing*, 20, pp. 2,347-66.

Stehman, S.V. (1999c), 'Estimating the Kappa Coefficient and its Variance under Stratified Random Sampling', *Photogrammetric Engineering and Remote Sensing*, 62, pp. 402-7.

Stehman, S.V. (1997a), 'Estimating Standard Errors of Accuracy Assessment Statistics under Cluster Sampling', *Remote Sensing of the Environment*, 60, pp. 258-69.

Stehman, S.V. (1997b), 'Selecting and Interpreting Measures of Thematic Classification Accuracy', *Remote Sensing of the Environment*, 62, pp. 77-89.

Stehman, S.V. (1996a), 'Use of Auxiliary Data to Improve the Precision of Estimators of Thematic Map Accuracy', *Remote Sensing Environment*, 58, pp. 169-76.

Stehman, S.V. (1996b), 'Cost-effective, Practical Sampling Strategies for Accuracy Assessment of Large-area Thematic Maps', in *Spatial Accuracy Assessment in Natural Resources and Environmental Sciences : Second International Symposium*, General Technical Report RM-GTR-277, Rocky Mountain Forest and Range Experiment Station, Fort Collins, Co.

Thompson, S.K. (1992), *Sampling*, New York: John Wiley.

Thompson, S.K. (1982), 'Adaptive Sampling of Spatial Point Processes', Ph. D. Thesis, Oregon State University.

Thompson, S.K. and G.A.F. Seber (1996), *Adaptive Sampling*, New York: John Wiley & Sons. Inc.

Tran, L.T. and L. Duckstein (2000), 'Comparison of Fuzzy Numbers using a Fuzzy Distance Measure', Manuscript.

Wickham, J.D., R.V. O'Neill, K.H. Ritters, T.G. Wade and K.B. Jones (1997), 'Sensitivity of Selected Landscape Pattern Metrics to Land cover Misclassification and Difference in Land cover Composition', *Photogrammetric Enginnering and Remote Sensing*, 63, pp. 397-402.

Winkler, G. (1995), *Image Analysis, Random Fields and Dynamic Monte Carlo Methods : A Mathematical Introduction*, New York: Springer.

Zeleny, M. (1982), *Multiple Criteria Decision Making*, New York : McGraw-Hill.

Zimmerman, H.J. (1987), *Fuzzy Sets, Decision Making and Expert Systems*, Boston: Kluwer Academic Publishers.

8

Global Warming and Climate Change

An Indian Perspective on Observations and Model Projections

K. Krishna Kumar • K. Rupa Kumar • D.R. Kothawale

In the context of the global warming scenario, understanding the regional climatic change over India is very important. These changes are expected to be manifest in the form of changes in seasonal temperature over the monsoon region which may induce a net change in monsoon rainfall, changes in its interannual to decadal scale variability and changes in the frequency and intensity of tropical cyclonic disturbances, jet streams, and changes in the boundaries of the arid and semi-arid zones in the region. Therefore, an attempt is made here to provide an assessment of the long-term changes in the monsoon rainfall and seasonal temperatures over India, using more than 100 years of instrumental records. Studies based on observed meteorological data during the last several decades have indicated clearly that the monsoon rainfall is trendless and is mainly random in nature over the period of instrumental record (1871–1998). However, on a sub regional scale, there do exist some areas with significant increasing and decreasing trends in monsoon rainfall. Surface air temperature over India shows a statistically significant increasing

---

The authors wish to thank G.B. Pant, Director, Indian Institute of Tropical Meteorology, Pune, for providing all necessary facilities to carry out this work. The meteorological data utilized in this work was obtained from the India Meteorological Department, Pune. We thank R.G. Ashrit for his useful inputs on future climate scenarios.

trend over the past century, amounting to about 0.4^oC per 100 years on an all-India scale. This warming has been contributed mainly by the winter and post-monsoon seasons. Further, the warming over India has been due mainly to an increase of the maximum temperatures rather than of the minimum temperatures observed over many other parts of the world. The examination of temperature data from a good number of industrial cities in India suggests that no generalization can be made regarding the effects of industrialization and urbanization on the ambient temperature changes.

1. Introduction

The last couple of decades have witnessed an unprecedented spurt in global attention, in the general public as well as in the scientific community, on various issues concerned with the climate and its variability – both natural and anthropogenic. The Intergovernmental Panel on Climate Change (IPCC), the International Geosphere-Biosphere Programme (IGBP), PAst Global changES (PAGES) and the Indian Climate Research Programme (ICRP) are some of the international and national scientific panels and programmes initiated to understand climate variability on different spatio-temporal scales aimed at developing global/regional scenarios of climate change and its impact.

The regional climatic impacts associated with global climatic change and their assessment are very important as agriculture, water resources, ecology, etc. are all vulnerable to climatic changes on the regional scale. This is particularly true for a country like India where the rainfall is highly seasonal with nearly 70-90 percent of its annual rainfall being received during in just four months (June-September) during the south-west monsoon season. Therefore, significant long-term changes, if any, in the monsoon rainfall could have adverse impacts on the large population living in this region. Rapid urbanization and industrialization in India also raise many questions as to whether this can result in significant temperature changes. These are some of the questions that we intend to address here using a large network of long-period meteorological data. Many studies in the past have addressed this problem but we try here to summarize these studies by updating the analysis with up to date meteorological data wherever possible. Also, the future scenarios of climate change over India, through coupled ocean-atmosphere model simulations, will also be addressed briefly.

2. Monsoon Rainfall Variability and Change

The rainfall over India varies on all time scales ranging from within a day to a decade, century and even longer. Apart from this, the rainfall also varies considerably on various spatial scales. Though the variability associated with each time scale is of importance in some respect or the other, year-to-year fluctuations and long-term trends, if any, are of great significance in making mitigatory efforts on the short-term and planning long-term strategies respectively. In view of this, a historical perspective on the rainfall observations and the variability of monsoon rainfall on the inter-annual time scale is also presented here before the issues related to long-term changes are addressed.

2.1. General Features of Monsoon Rainfall

The south-west monsoon season from June to September accounts for about 70-90 percent of the annual rainfall over a major part of India. Therefore, the failure of the monsoon in any particular year over a large area results in drought, ultimately affecting the regional and national economies. Rainfall, being one of the most important and conspicuous of all atmospheric processes and having direct relevance to the very survival of all human beings, plants and animals, has always attracted the greatest attention of natural philosophers and meteorologists. Therefore, the recording of rainfall, perhaps the first meteorological parameter to be measured, has a long history. Continuous records of the south-west monsoon rainfall are available for the Indian cities of Madras and Bombay since 1813 and 1817 respectively. There are about 10 stations in India which have more than 150 years of continuous rainfall records for the monsoon season. A reasonably good network of stations of climatological significance, with at least one rain-gauge per district having a continuous and homogenous rainfall record, has been available only since 1871 (Parthasarthy *et al.* 1987). A network of 306 stations, one from each of the districts in the plain contiguous region of India and distributed uniformly over the country, has been selected (Mooley and Parthasarthy 1984a). Figure 8.1 shows the rain-gauge network considered over India. In view of the limitations of the areal representation of rain-gauges in hilly areas and also due to their sparsity, the rainfall data from the hilly regions in the Himalayan mountain ranges, comprising Jammu and Kashmir, Himachal Pradesh, the hills of west Uttar Pradesh, Sikkim, parts of Sub-Himalayan West Bengal, Arunachal Pradesh and parts of north Assam have not been considered. The total area of India is 3.292×10^6 km^2, and the area considered measures 2.88×10^6 km^2, amounting to about 90% of the total area of the country. This indicates that the exclusion of hilly areas does not affect the macro-regional means of rainfall over the country.

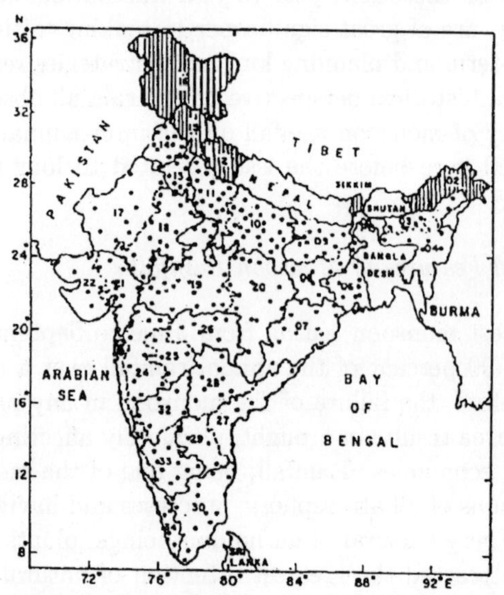

Figure 8.1: Network of 306 Raingauge Stations and Demarcation of 29 Meteorological Subdivisions Considered in the Study. The Sub divisions of 1,2,12,15,16 and 35 are not considered.

The quantity, All India Summer Monsoon Rainfall (AISMR) has been calculated by averaging arithmetically the district area weighted rainfall values of 306 stations. The average area represented by each rain-gauge works out to be 100 × 100 km, but in individual cases, it varies from 70 × 70 km in UP to 150 × 150 km in west Rajasthan. Figure 8.1 also shows the latest meteorological subdivisions into which the country has been divided and the area considered in the preparation of the monsoon rainfall series. As can be seen from the figure, the four hilly subdivisions (whose numbers are 2, 12, 15 and 16) and two subdivisions far away from the mainland, that is, Bay Islands (Subdivision No.1) and Arabian Sea Islands (Subdivision No. 35) have not been considered while quantifying the AISMR as well as the subdivisional summer monsoon rainfall series. The summer monsoon area-weighted rainfall series for each of the remaining 29 meteorological subdivisions have been prepared by assigning the district areas as the weight for each rain-gauge station.

Some important statistical characteristics of all of India as well as 29 meteorological subdivisions have been computed utilizing the data for the period 1871-1990. In Table 8.1 mean (mm), standard deviation (mm), coefficient of variation (CV, percent), auto correlation coefficients for lags 1, 2 and 3, the number of droughts and floods as well as the years of lowest and highest rainfall with corresponding rainfall during the period 1871-1990 are recorded. All India summer monsoon rainfall has a mean of 852.6 mm with a CV of 9.9 percent. Among the subdivisions, coastal Karnataka receives the highest rainfall of 2,852 mm followed by Konkan and Goa and Sub-Himalayan West Bengal. West Rajasthan receives the lowest rainfall of 255 mm, followed by Tamil Nadu and Rayalaseema. While the highest CV is found over Saurashtra and Kutch (44.4 percent) followed by West Rajasthan (39.1 percent) and Punjab (35.2 percent), the lowest CV is found over north Assam (11.26 percent) followed by south Assam and Orissa. The lag correlations are not statistically significant for any of the rainfall series, indicating that there is no persistence in these series. The years of highest and lowest rainfall for all India are 1961 (1,020.5) and 1877 (604.4) with a range of 416 mm which is about 48 percent of the long-term mean. The years of highest and lowest rainfall at individual subdivisions are quite different from those of all of India. Though there is some spatial coherence in the years of lowest rainfall on a subdivisional scale indicating large-scale droughts, the years of highest rainfall seem to be highly region-specific, presumably due to some local events contributing to the high rainfall over a particular subdivision.

2.2. Droughts and Floods

Though the Indian summer monsoon is a very stable and dependable source of water for the country, there are considerable year-to-year changes superimposed on this stable picture on an all-India scale. This can be seen in the All-India Summer Monsoon Rainfall anomaly series during the period 1871-1998 presented in Figure 8.2. Years with rainfall anomalies exceeding one standard deviation on the positive/negative sides are categorized as flood/drought years respectively. There are, in all, 22 drought years and 18 flood years

Table 8.1

Summer Monsoon Rainfall Statistics of all-India, Subdivisions and Homogeneous Regions of India during 1871-1990

S. No.	Sub-division Name	Mean	Std. Dev.	C.V. %	Auto Correlation			Number of Years		Rainfall (in mm)		Rainfall (in mm)	
					Lag-1	Lag-2	Lag-3	Drought	Flood	Lowest	Year	Highest	Year
1	North Assam	1453.54	163.62	11.26	0.05	-0.19	0.18	23	17	1015.2	1884	2053.5	1993
2	South Assam	1449.08	182.71	12.61	-0.02	-0.03	0.08	23	22	1031.6	1980	1891.6	1918
3	Sub-Himalayan W. Bengal	2001.16	298.88	14.94	0.01	-0.11	0.01	20	21	1205.0	1891	2738.5	1890
4	Gangetic West Bengal	1145.18	167.14	14.60	-0.05	-0.14	0.08	22	24	837.3	1892	1601.2	1993
5	Orissa	1166.66	157.87	13.53	0.07	-0.02	0.17	18	19	755.0	1974	1583.4	1956
6	Bihar Plateau	1097.29	157.75	14.38	0.03	0.07	0.08	21	23	695.4	1966	1505.4	1942
7	Bihar Plains	1037.08	198.86	19.17	0.01	-0.13	0.02	19	18	538.7	1908	1584.5	1987
8	East Uttar Pradesh	909.93	204.28	22.45	0.11	-0.04	-0.16	18	16	309.4	1877	1433.4	1980
9	West Uttar Pradesh Plains	767.42	179.29	23.36	-0.06	0.02	0.09	17	17	164.6	1877	1131.1	1936
10	Haryana	456.62	141.60	31.01	-0.04	0.05	.14	22	21	142.4	1987	896.5	1933
11	Punjab	493.63	173.58	35.16	-0.09	0.08	0.10	15	18	151.2	1987	1012.0	1971
12	West Rajasthan	255.62	99.94	39.10	0.05	0.00	0.00	14	18	33.6	1918	573.2	1908
13	East Rajasthan	635.97	167.60	26.35	-0.03	0.11	0.08	20	19	194.5	1877	1118.5	1917
14	West Madhya Pradesh	918.54	166.10	18.08	0.00	0.25	0.11	21	20	519.2	1918	1309.3	1973
15	East Madhya Pradesh	1197.50	192.13	16.04	-0.03	0.11	0.15	20	18	708.2	1979	1836.8	1884
16	Gujarat	863.21	270.58	31.35	-0.04	-.03	0.00	21	19	211.0	1899	1577.7	1878
17	Saurashtra & Kutch	432.12	191.89	44.41	-0.07	-0.04	0.08	17	19	62.5	1987	1079.5	1878
18	Konkan & Goa	2385.00	470.54	19.73	0.06	0.02	-0.04	14	17	1051.9	1918	3754.4	1878
19	Madhya Maharashtra	579.49	123.23	21.26	0.06	-0.11	-0.19	19	17	210.5	1918	941.6	1878
20	Marathwada	695.21	196.56	28.27	0.13	-0.03	-0.03	18	22	243.5	1918	1300.6	1892
21	Vidarbha	950.37	185.94	19.56	-0.13	0.17	0.06	20	20	383.3	1899	1371.7	1883
22	Coastal Andhra Pradesh	506.90	112.30	22.16	-0.10	0.07	0.15	18	20	308.5	1888	780.3	1978
23	Telengana	722.22	168.69	23.36	0.07	0.08	0.09	21	22	371.4	1877	1186.0	1988
24	Rayalaseema	422.09	120.92	28.65	-0.11	-0.17	0.06	22	19	192.4	1904	791.0	1878
25	Tamil Nadu	309.31	69.92	22.61	-0.09	-0.02	0.00	19	20	169.8	1918	466.3	1910
26	Coastal Karnataka	2852.24	507.28	17.79	-0.12	-0.01	-0.09	16	22	1601.7	1918	4623.1	1975
27	North Interior Karnataka	600.92	119.16	19.83	0.00	-0.07	0.13	16	16	340.1	1918	904.4	1914
28	South Interior Karnataka	503.29	101.54	20.18	-0.15	-0.20	0.13	21	20	264.6	1918	776.3	1897
29	Kerala	1938.29	372.85	19.24	0.04	-0.09	0.00	20	19	1150.4	1918	3115.4	1924
30	All-India	852.60	84.34	9.89	-0.12	0.04	0.08	22	18	604.4	1877	1020.5	1961
31	Homogeneous India	757.37	118.47	15.64	-0.10	0.15	0.09	21	20	385.7	1899	986.3	1961
32	Core-Monsoon India	858.39	147.05	17.13	-0.06	0.14	0.08	18	23	414.8	1899	1135.4	1892
33	North-west India	490.25	131.81	26.89	-0.08	0.02	0.09	20	22	162.4	1899	817.4	1917
34	West Central India	933.36	125.39	13.43	-0.07	0.18	0.05	16	22	532.8	1899	1211.7	1961
35	Central North-west	1002.63	112.30	11.20	0.01	-0.04	0.01	20	16	614.8	1877	1354.0	1936
36	North-east India	1419.41	120.82	8.51	0.07	-0.20	0.05	22	17	1137.1	1992	1793.4	1918
37	Peninsular India	659.61	97.86	14.84	-0.18	-0.16	0.02	16	21	405.0	1918	938.2	1878

Table 8.2

Occurrence of Droughts and Floods over 29 Meterological Subdivisions in India: 1871-1994

Year	1	2	3	4	5	6	7	8	9	10	11	12	13	14	15	16	17	18	19	20	21	22	23	24	25	26	27	28	29
1871	•			•	○	•	•	•	•					•				○	○	○	○		○				○		
1872	•			○	•	○				•										•			○			•			•
1873	○	○	○	○	○		○																○		○	○	○	○	
1874				○				•	•					•	•			•	•					•		•	•	•	
1875		•								•	•			•					•	•			○					○	
1876																		○	○	○			○	○			○	○	
1877	○		○		○	○	○	○	○		○	○	○	○	○		○		○	○	○	○			○				
1878	•	•	•		○										○	•	•	•	•		•	•	•	•		•	•		•
1879	•	•	•	○			•	•	•				•																
1880		•		•	•			○														○							
1881				•															○				○			○		○	○
1882	○																		•	•							•	•	•
1883	○				•			○	○	○					○				•	•	•		•						
1884	○	○	○			○	○		•			•	•	•	•				•			○	○			○			
1885				•	○	•			•	•	○																		•
1886	•			•	○	•	•	•	•					•				○	○	○	○		○				○		
1887	•			○	•	○				•										•			○			•			•
1888	○	○	○	○	○		○																○		○	○	○	○	
1889				○				•	•					•	•			•	•					•		•	•	•	
1890		•								•	•			•					•	•			○					○	
1891																		○	○	○			○	○			○	○	
1892	○		○		○	○	○	○	○		○	○	○	○	○		○		○	○	○	○			○				
1893	•	•	•		○										○	•	•	•	•		•	•	•	•		•	•		•
1894	•	•	•	○			•	•	•				•																
1895		•		•	•			○														○							
1896				•															○				○			○		○	○
1897	○																		•	•							•	•	•
1898	○				•			○	○	○					○				•	•	•		•						
1899	○	○	○			○	○		•			•	•	•	•				•			○	○			○			
1900				•	○	•			•	•	○																		•
1901	•			•	○	•	•	•	•					•				○	○	○	○		○				○		
1902	•			○	•	○				•										•			○			•			•
1903	○	○	○	○	○		○																○		○	○	○	○	
1904				○				•	•					•	•			•	•					•		•	•	•	
1905		•								•	•			•					•	•			○					○	
1906																		○	○	○			○	○			○	○	
1907	○		○		○	○	○	○	○		○	○	○	○	○		○		○	○	○	○			○				
1908	•	•	•		○										○	•	•	•	•		•	•	•	•		•	•		•
1909	•	•	•	○			•	•	•				•																
1910		•		•	•			○														○							

○ Drought Year • Flood Year

Table 8.2: Contd.

Year	1	2	3	4	5	6	7	8	9	10	11	12	13	14	15	16	17	18	19	20	21	22	23	24	25	26	27	28	29
1911	●			●	○	●	●	●	●					●				○	○	○	○		○				○		
1912	●			○	●	○				●										●			○			●			●
1913	○	○	○	○	○		○																○		○	○	○	○	
1914				○				●	●					●	●			●	●					●		●	●	●	
1915		●								●	●			●					●	●			○					○	
1916																		○	○	○			○	○			○	○	
1917	○		○		○	○	○	○	○		○	○	○	○	○		○		○	○	○	○			○				
1918	●	●	●		○										○	●	●	●	●		●	●	●	●		●	●		●
1919	●	●	●	○			●	●	●				●																
1920		●		●	●			○														○							
1921				●															○				○			○		○	○
1922	○																		●	●							●	●	●
1923	○				●			○	○	○					○				●	●	●		●						
1924	○	○	○			○	○		●			●	●	●	●				●			○	○			○			
1925				●	○	●			●	●	○																		●
1926	●			●	○	●	●	●	●					●				○	○	○	○		○				○		
1927	●			○	●	○				●										●			○			●			●
1928	○	○	○	○	○		○																○		○	○	○	○	
1929				○				●	●					●	●			●	●					●		●	●	●	
1930		●								●	●			●					●	●			○					○	
1931																		○	○	○			○	○			○	○	
1932	○		○		○	○	○	○	○		○	○	○	○	○		○		○	○	○	○			○				
1933	●	●	●		○										○	●	●	●	●		●	●	●	●		●	●		●
1934	●	●	●	○			●	●	●				●																
1935		●		●	●			○														○							
1936				●															○				○			○		○	○
1937	○																		●	●							●	●	●
1938	○				●			○	○	○					○				●	●	●		●						
1939	○	○	○			○	○		●			●	●	●	●				●			○	○			○			
1940				●	○	●			●	●	○																		●
1941	●			●	○	●	●	●	●					●				○	○	○	○		○				○		
1942	●			○	●	○				●										●			○			●			●
1943	○	○	○	○	○		○																○		○	○	○	○	
1944				○				●	●					●	●			●	●					●		●	●	●	
1945		●								●	●			●					●	●			○					○	
1946																		○	○	○			○	○			○	○	
1947	○		○		○	○	○	○	○		○	○	○	○	○		○		○	○	○	○			○				
1948	●	●	●		○										○	●	●	●	●		●	●	●	●		●	●		●
1949	●	●	●	○			●	●	●				●																
1950		●		●	●			○														○							

○ Drought Year ● Flood Year

Table 8.2: Contd.

Year	1	2	3	4	5	6	7	8	9	10	11	12	13	14	15	16	17	18	19	20	21	22	23	24	25	26	27	28	29
1951	●			●	○	●	●	●	●					●				○	○	○	○		○				○		
1952	●			○	●	○				●										●			○			●			●
1953	○	○	○	○	○		○																○		○	○	○	○	
1954				○				●	●					●	●			●	●					●		●	●	●	
1955		●								●	●			●					●	●			○					○	
1956																		○	○	○			○	○			○	○	
1957	○		○		○	○	○	○	○		○	○	○	○	○		○		○	○	○	○			○				
1958	●	●	●		○										○	●	●	●	●		●	●	●	●		●	●		●
1959	●	●	●	○			●	●	●				●																
1960		●		●	●			○														○							
1961				●															○				○			○		○	○
1962	○																		●	●							●	●	●
1963	○				●			○	○	○					○				●	●	●		●						
1964	○	○	○			○	○		●			●	●	●	●				●			○	○			○			
1965				●	○	●			●	●	○																		●
1966	●			●	○	●	●	●	●					●				○	○	○	○		○				○		
1967	●			○	●	○				●										●			○			●			●
1968	○	○	○	○	○		○																○		○	○	○	○	
1969				○				●	●					●	●			●	●					●		●	●	●	
1970		●								●	●			●					●	●			○					○	
1971																		○	○	○			○	○			○	○	
1972	○		○		○	○	○	○	○		○	○	○	○	○		○		○	○	○	○			○				
1973	●	●	●		○										○	●	●	●	●		●	●	●	●		●	●		●
1974	●	●	●	○			●	●	●				●																
1975		●		●	●			○														○							
1976				●															○				○			○		○	○
1977	○																		●	●							●	●	●
1978	○				●			○	○	○					○				●	●	●		●						
1979	○	○	○			○	○		●			●	●	●	●				●			○	○			○			
1980				●	○	●			●	●	○																		●
1981	●			●	○	●	●	●	●					●				○	○	○	○		○				○		
1982	●			○	●	○				●										●			○			●			●
1983	○	○	○	○	○		○																○		○	○	○	○	
1984				○				●	●					●	●			●	●					●		●	●	●	
1985		●								●	●			●					●	●			○					○	
1986																		○	○	○			○	○			○	○	
1987	○		○		○	○	○	○	○		○	○	○	○	○		○		○	○	○	○			○				
1988	●	●	●		○										○	●	●	●	●		●	●	●	●		●	●		●
1989	●	●	●	○			●	●	●				●																
1990		●		●	●			○														○							
1991	○		○		○	○	○	○	○		○	○	○	○	○		○		○	○	○	○			○				
1992	●	●	●		○										○	●	●	●	●		●	●	●	●		●	●		●
1993	●	●	●	○			●	●	●				●																
1994		●		●	●			○														○							
Drought	23	23	20	22	18	21	19	18	17	22	15	14	20	21	20	21	17	14	19	18	20	18	21	22	19	16	16	21	20
Flood	17	22	21	24	19	23	18	16	17	21	18	18	19	20	18	19	19	17	17	22	20	20	22	19	20	22	16	20	19

on an all-India basis. Similar statistics (during 1871-1995) for all the subdivisions are presented in Table 8.1. Contrary to general belief, the high rainfall subdivisions in northeast India have more drought years than those over north-west and west central India. Also, the number of flood years over north-east India is less than that over north-west and west central India.

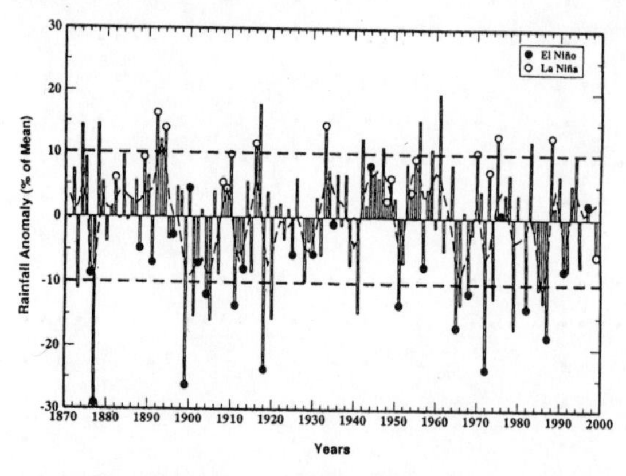

Figure 8.2: Variation of Monsoon Rainfall Anomaly (percentage of mean) of All India During 1871-1999

In view of the high spatial variability of monsoon rainfall over the country, it would be of interest and importance to examine the spatial extents of the floods and droughts delineated above. For this purpose, yearwise (1871-1994) occurrences of floods and droughts over 29 meteorological subdivisions of India are presented in Table 8.2. The areal extents of some of the large-scale droughts and floods are listed in Table 8.3.

2.3. Interannual Variability

Several global climatic factors have been identified for the interannual (year-to-year) variability of monsoon rainfall over India. Prominent among these is the phenomenon known as El Niño-Southern Oscillation (ENSO). While El Niño is a warming of the surface waters of the tropical eastern Pacific Ocean, the southern oscillation (SO) refers to a large-scale see-saw exchange in sea level air pressure between areas of the western and the south-eastern Pacific. Gilbert Walker (1924) was the first to demonstrate the possible connection between the SO and the monsoon rainfall over India. The profound impact of this coupled ocean - atmospheric phenomenon can be understood from the coincidence of most of the major droughts and floods in India with the occurrence of El Niño (warm ENSO phase) and La Niña (cold ENSO phase) events in the Pacific as shown in Figure 8.2 with filled and open circles respectively. During this period, 11 out of a total of 22 drought years were El Niño years, whereas only 2 out of 18 excess rainfall years were El Niño years. More information on the global long-distance connections of monsoon rainfall and prediction can be had from Krishna Kumar *et al.* (1995, 1999); and Pant and Rupa Kumar (1997).

Table 8.3
Areal Extent (Percentage Area of the Country) of Some of the Large-scale (28 percent to 73 percent Area) Droughts and Floods over India during 1871-1994

Drought		Flood	
Year	Area (%)	Year	Area (%)
1877	66.8	1892	46.7
1899	73.0	1917	52.8
1918	68.2	1956	28.8
1972	49.5	1961	48.9
1979	49.2	1988	48.9
1987	64.3		

These tables 8.1–8.3 also provide a bird's eye view of the frequency of flood/drought situations over different subdivisions of India.

2.4. *Low-frequency Variations*

In Figure 8.2, the 5-point Gaussian low-pass filtered values of all India rainfall anomalies are also provided to indicate its low frequency behaviour. The low-pass filtered values of rainfall anomalies are generally negative during the periods 1896-1930 and 1962-1987 and positive during 1874-1895 and 1942-1961. Apart from these multidecadal wet and dry spells, there are also some above-normal and below-normal rainfall periods spanning over less than 10 years.

2.5. *Long-term Trends*

Long-term trends in the monsoon rainfall over India are also important, particularly in the arid and semi-arid regions where any significant long-term climatic change will have a far-reaching impact on society. The long-term changes in Indian rainfall have been examined by several workers in India. Reviews of such earlier studies are given by Mooley and Parthasarthy (1984b), Pant *et al.* (1993), Pant *et al.* (1991), Pant *et al.* (1988), Parthasarthy and Dhar (1978), Parthasarthy *et al.* (1993), Rupa Kumar *et al.*, (1992), Sarker and Thapliyal (1988), Soman *et al.* (1988), Srivastava *et al.* (1992) and Thapliyal and Kulshrestha (1991). These studies ranged from the long-term trends at individual stations to those in the all India mean rainfall series. However, most of the studies have clearly pointed out that the monsoon rainfall is trendless and is mainly random in nature over a long period of time, particularly on all India scale (Mooley and Parthasarthy, 1984a). The trendlessness of monsoon rainfall on the all India scale can be visualized from the time series of AISMR shown in Figure 8.2.

Authors Rupa Kumar *et al.* (1992) provide a comprehensive picture of long-term trends on smaller regional scales by examining the linear trends in Indian summer monsoon rainfall using data at 306 stations during 1871-1984. In the present study, trends in the monsoon rainfall are presented by updating the rainfall data till 1990 of 306 stations. The spatial variation of the linear trend in monsoon rainfall, expressed as a percentage of the long-term mean per 100 years, is shown in Figure 8.3. It can be seen from the figure that the areas with an increasing trend in the monsoon seasonal rainfall are found along the west coast, north Andhra Pradesh and north-west India and those with a decreasing trend, over east Madhya Pradesh and adjoining areas, north-east India and parts of Gujarat and Kerala.

3. Long-term Trends in Surface Air Temperature over India

The monitoring and analysis of atmospheric temperatures on the global and regional scales has acquired special importance in the last few decades due to the clear indications of global warming in the post-industrial era. It has been concluded in the latest report from the Intergovernmental Panel on Climate Change (IPCC 2001) (Houghton 1992, 1990) that the global mean surface air temperature has increased by 0.6^oC during the 20th century, with many

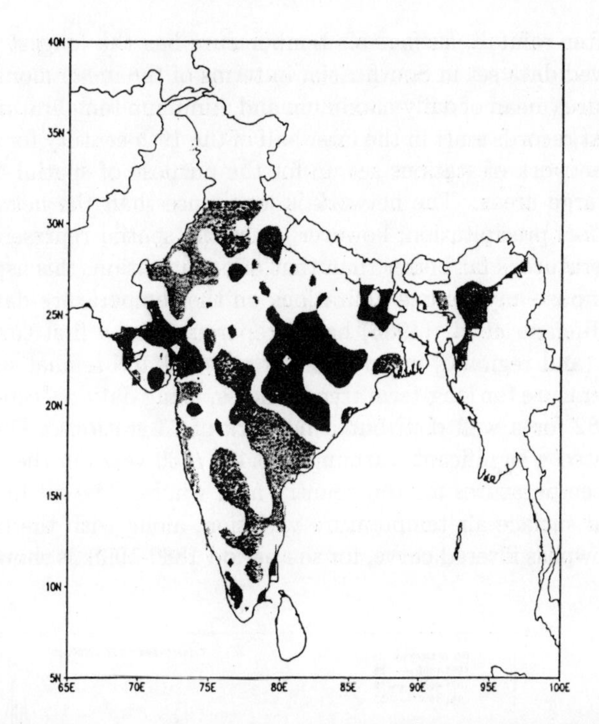

Figure 8.3: Spatial Patterns of Linear Trends in Summer Monsoon Rainfall during 1871–1990. Shaded areas Indicate Negative (Dark) and Positive (Light) Trends with Magnitudes Greater than 5% of the Mean per 100 years. Dotted areas Indicate Trends Significant at 5% Level.

occurring in the last decade or so. In particular, the variability of surface temperature on the regional and global scales has come to be viewed with as much importance as the monitoring of precipitation. The significant increase in the mean annual global surface air temperature during the past century, predominantly over the northern hemisphere, is probably the most widely quoted aspect of contemporary climatic change. Climatic changes over the Asian monsoon region, of which India is a major component, are of global concern, particularly in view of the high concentration of human population. We have already observed in the previous section that the Indian summer monsoon rainfall, on which the country's economy is critically dependent, has shown remarkable stability over the past 120 years, at least on a countrywide scale. While this can be viewed as a reassuring facet of climatic change in this region, enough indications are available regarding South Asia's share of global warming.

After rainfall, surface air temperature has the longest and best achieved data set in South Asia, in terms of the mean monthly temperature (mean of daily maximum and minimum temperatures). The earliest records start in the later half of the 19th century for a reasonable network of stations set up for the purpose of spatial averaging over large areas. The network is less dense than the network that measures precipitation; however, since the spatial representation of temperature is far better than that of precipitation, this aspect does not impose any serious limitations on the temperature data analysis. Hingane *et al.* (1985) have prepared, for the first time, an all India (and regional) mean series of seasonal and annual surface air temperature for long-term trend studies, using data collected during 1901-82 for a well distributed network of 73 stations. Their study indicated a significant warming of $0.4^oC/100$ years in the mean annual temperatures for the country as a whole. The all India mean annual surface air temperature variation, along with the trend line and lowpass filtered curve, for the period 1880-1998, is shown Figure 8.4.

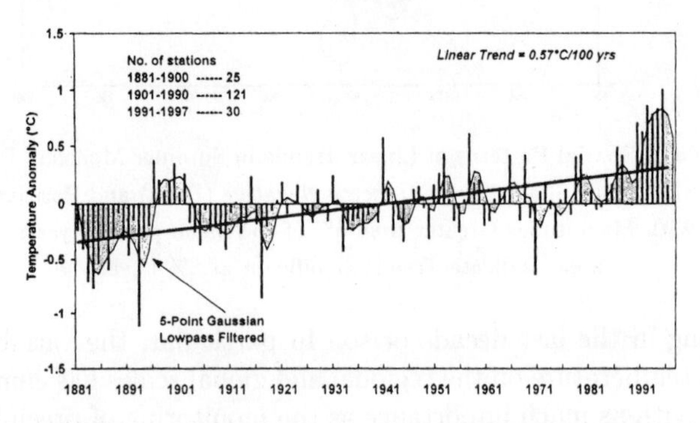

Figure 8.4: Variation of Annual Surface Air Temperature Anomaly (^oC) of All-India during 1991-97.

This figure (8.4) is an extended version of that presented by Hingane *et al.* (1985); however, the number of stations is less and variable (depending upon the available network) before 1901 and after 1982 as indicated in the figure. The steep increase in temperatures in the recent decades may be viewed with some caution owing to the less number of stations utilized in computing the all India average for that period. The spatial distribution of the linear trend (in ^oC per 100 years) with an updated data during 1901-1990 is shown in Figure 8.5. There are large areas along the west coast that show a significant warming trend excepting the middle portion, the interior peninsula and over north-east India. North-west India is conspicuous by its significant cooling trend. Though the pattern of surface temperature change displays some cellular areas of warming and cooling, it can be seen that there are broad areas of regional-scale warming over a major part of the country. Another study of Srivastava *et al.* (1992) on the decadal trends of temperature over India using a large network of stations also confirms the above described nature of temperature trends over India.

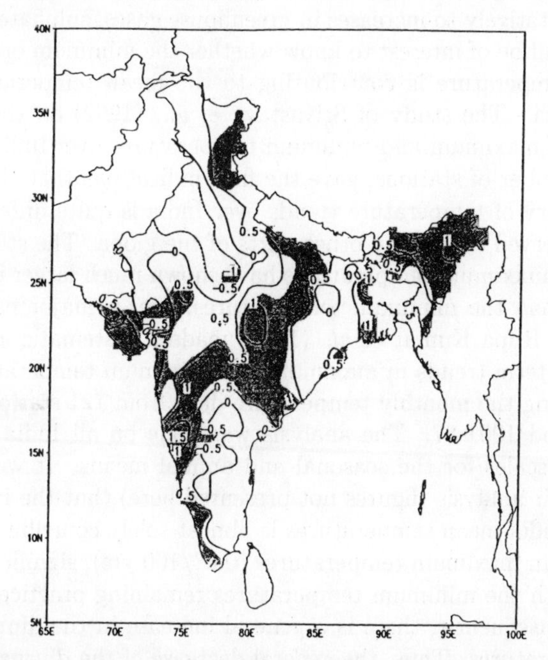

Figure 8.5: Spatial Patterns of Linear Trends in Mean Annual Surface Air Temperature over India (in ^oC per 100 Years) during 1901-90. Shading Indicates Significance at 5% Level.

While significant warming in the mean temperatures is a matter of concern, greater attention is now being paid to the manifestations of such warming in terms of the day-time (represented by the daily maximum) and night-time (represented by the daily minimum) temperatures. It is believed that such diurnal asymmetries in temperature trends, if any, have close links with the changes in cloudiness, humidity, atmospheric circulation patterns, winds and soil moisture, which are likely to be significantly largely over land areas (Houghton *et al.*, 1990). Karl *et al.* (1993) made a systematic analysis of the maximum and minimum temperatures over several land areas (USA, former USSR, China, Canada, Japan, South Africa, Australia, etc.) during 1951-90 and reported that the minimum temperatures have generally increased more rapidly than the maximum temperatures, leading to a decrease in the diurnal temperature range. They attributed the decrease in the diurnal temperature range to the possible increases in cloud cover and linked them tentatively to increases in greenhouse gases, sulphate aerosols, etc. It will be of interest to know whether the minimum or the maximum temperature is contributing to the mean temperature raise over India. The study of Srivastava *et al.* (1992) on the decadal trends of maximum and minimum temperatures over India, using a large number of stations, gave the first indications that the diurnal asymmetry of temperature trends over India is quite different from that observed over many other parts of the globe. The study found that the maximum temperatures have shown much larger increasing trends than the minimum temperatures, over a major part of the country. Rupa Kumar *et al.* (1994) made a systematic analysis of the long-term trends in maximum and minimum temperatures over India using the monthly temperature data from 121 stations during the period 1901-87. The analysis was done on all India and sub-regional scales for the seasonal and annual means. It was evident from their analysis (figures not presented here) that the increase in the all-India mean temperatures is almost solely contributed by the increase in maximum temperatures (0.6^o/100 yrs), significant at 1% level, with the minimum temperatures remaining practically trendless. Consequently, there is a general increase in the diurnal range of temperatures. Thus, the general decrease of the diurnal range of temperatures observed by Karl *et al.* (1993) over large land areas of the globe has only a limited presence in India, confined to the interior peninsula (spatial pattern of trends not shown here).

4. Precipitation and Surface Temperature Trend in Major Cities in India

It is difficult to interpret the observed temperature changes in terms of cause and effect. It is likely that some of the changes that have taken place are part of variations on a global scale. It is known that intense deforestation activities have taken place along the foothills

Table 8.4:
Linear trend (^oC/100 years) in Mean Temperature and Monsoon Rainfall (mm/year)

Sr.	Station	Temp. for DJF			MAM			JJAS		
		Mean	SD	Trend	Mean	SD	Trend	Mean	SD	Trend
1	All-India	20.0	0.4	0.6*	28.1	0.5	0.3	27.6	0.3	
2	New Delhi	15.6	0.7	-0.3	28.1	1.1	-0.9*	31.1	0.7	-0.4
3	Jodhpur	18.6	0.8	-0.1	29.9	1.0	0.1	31.1	0.7	0.2
4	Ahmedabad	22.0	1.0	-1.6**	31.0	0.7	-0.2	29.9	0.7	-0.2
5	Indore	18.9	0.7	0.5*	28.8	0.7	0.7**	26.7	0.5	0.5*
6	Sagar	18.9	0.7	0.4	29.7	1.0	0.2	27.4	0.6	0.1
7	Gauhati	17.9	1.3	2.7*	25.1	0.9	1.4**	28.1	1.1	-2.1
8	Kolkata	21.0	0.7	1.6**	29.5	0.8	0.7*	29.3	0.4	0.7**
9	Bikaner	16.0	1.0	-0.2	29.3	1.0	-0.2	32.8	0.7	-0.2
10	Nagpur	21.5	0.6	0.2	31.6	0.9	0.1	28.5	0.6	0.3
11	Mumbai	24.9	0.7	0.8**	28.5	0.5	0.4	27.9	0.3	0.3*
12	Pune	21.4	0.8	-0.7	28.3	0.6	-0.3	25.5	0.4	0.4**
13	Bangalore	21.8	0.6	0.5*	26.9	0.5	0.5*	23.8	0.4	0.4*
14	Hyderabad	22.7	0.8	0.1	30.8	0.8	-0.4	27.2	0.6	-0.3
15	Minambakkam	25.0	0.5	0.6**	30.2	0.6	1.4**	30.7	0.5	-0.2
16	Trivandrum	26.7	0.6	1.7**	28.4	0.5	0.9**	26.3	0.4	0.9**

Sr.	Station	ON			Annual Temp.			Monsoon		
		Mean	SD	Trend	Mean	SD	Trend	Mean	SD	Trend
1	All-India	24.0	0.5	0.6**	25.2	0.3	0.3*	847.8	80.1	0.2
2	New Delhi	23.3	0.9	-0.5	25.1	0.5	-0.5*	604.1	232.5	1.8*
3	Jodhpur	25.5	1.0	0.0	26.8	0.6	0.1	339.6	174.1	0.7
4	Ahmedabad	27.2	1.1	-1.6**	27.8	0.6	-0.8**	749.3	310.8	0.3
5	Indore	23.0	1.0	0.9*	24.6	0.5	0.6**	867.1	249.6	1.7*
6	Sagar	23.6	1.0	1.0	25.2	0.5	0.3	1114.8	314.7	0.8
7	Gauhati	23.8	0.5	0.3	24.1	0.3	0.4**	1073.7	212.0	0.8
8	Kolkata	25.9	0.7	1.6**	26.7	0.4	1.1**	1238.2	274.4	1.6
9	Bikaner	24.6	1.1	-0.1	26.4	0.6	-0.3	247.7	125.5	-0.1
10	Nagpur	24.5	0.8	0.3	26.9	0.4	0.2	1007.5	222.2	-2.0*
11	Mumbai	28.1	0.7	1.0**	27.3	0.3	0.6**	1855.2	517.5	5.2*
12	Pune	24.0	0.8	-0.3	24.9	0.4	-0.2	535.7	158.4	1.4*
13	Bangalore	22.7	0.5	0.5*	23.9	0.4	0.5**	510.0	161.7	0.4
14	Hyderabad	24.3	0.8	-0.3	26.5	0.5	-0.2	593.0	161.7	0.4
15	Minambakkam	27.1	0.4	0.4*	28.5	0.3	0.5**	399.4	151.3	1.3*
16	Trivandrum	26.5	0.6	1.4**	27.0	0.4	1.2**	856.1	256.3	-0.4

* Significant at 5% Level

Note: DJF stands for December, January and February; other column headings stand for other months of the year.

** Significant at 1% Level

of the Himalaya and Assam region and land-use patterns have undergone definite changes over parts of Rajasthan and Punjab (northwest India). Pant and Hingane (1988) found significant surface cooling associated with significant rainfall increase in the peripheral regions of the Rajasthan desert and proposed the increased area under irrigation as one of the main causal factors. Apart from this, understanding the role of rapid industrialization/urbanization in India on the observed climate changes will be of great value in planning mitigatory steps, wherever possible, to contain the changes in ambient temperature/rainfall variations. Rupa Kumar and Hingane (1988) have examined the trends in surface air temperatures over six major industrialized cities in India during the past century and found Mumbai, Kolkata and Bangalore to have a significant warming trend while Delhi showed a significant cooling trend. Here, we present an updated analysis on the trends in monsoon rainfall and surface air temperature during 1901-1998 at 16 selected cities. The time series of annual temperature and monsoon seasonal rainfall at these 16 stations along with the trend line is shown in Figure 8.6.

The results of trend analysis are summarized in Table 8.4. From the table and the figure it can be seen that in general, the trends in surface temperature are more pronounced in the post-monsoon and winter seasons as compared to the pre-monsoon and monsoon seasons. Out of the 16 stations, Kolkata, Mumbai, Bangalore, Indore, Minambakkam (Chennai) and Trivandrum have shown statistically significant increasing trends. But New Delhi and Ahmedabad have shown a significant decreasing trend. From a careful examination of these trends in conjunction with Figure 8.5, it appears as if the direction and significance of the trends at any given station is determined more by the location of the station with respect to the broad areas of increasing/decreasing trends observed over India (Fig. 8.5) rather than the station being highly industrialized/non-industrialized. In terms of rainfall changes, the most prominent increase in monsoon rainfall has occurred at Mumbai, Pune, Bangalore, New Delhi, Indore and Minambakkam. Kolkata also shows a considerable increase in the monsoon rainfall, though it is not statistically significant. On the other hand, Nagpur shows a significant decrease in monsoon rainfall. Therefore, it appears that the warming observed over India cannot be attributed fully to industrialization and urbanization, as neither is the warming trend confined to the industrial stations nor is the cooling trend confined to non-industrial stations. This suggests that no generalization can be made on the effect of industrialization

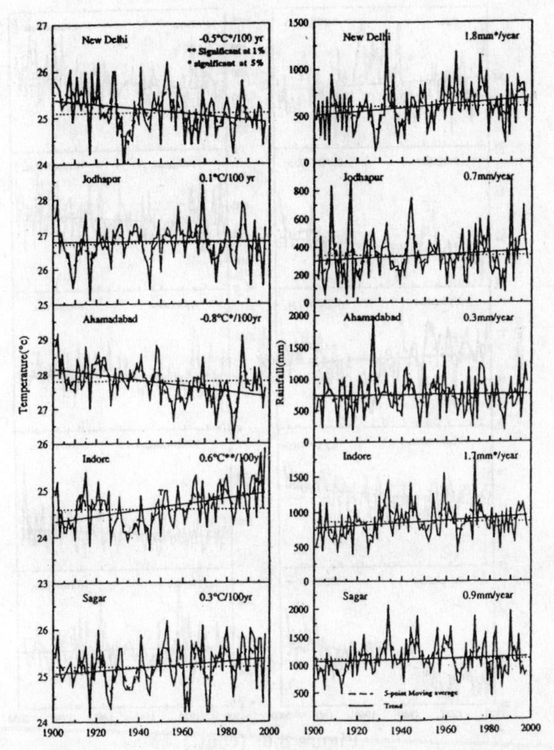

Figure 8.6: Variation of Mean Annual Temperature (^oC) and Summer Monsoon Rainfall (mm) at Major Cities in India

and urbanization on the ambient temperature/rainfall variations in India.

5. FUTURE CLIMATE SCENARIO OF INDIA

The transient climate change simulations of a coupled ocean - atmosphere general circulation model (ECHAM4/OPYC3 of MPI, Hamburg, Germany) suggest an increase by 13 percent in the mean monsoon rainfall over India by 2039 relative to 20th century climatology under the influence of the increasing concentration (1% per annum) of greenhouse gases. The increased mean is also accompanied by an increase in the interannual variability, suggesting an increased frequency of extreme drought/flood conditions in the scenario (Ashrit *et al.* 2001). This climate scenario also suggests a warming by 1.3^oC of the mean annual temperature by 2039 accompanied by maximum

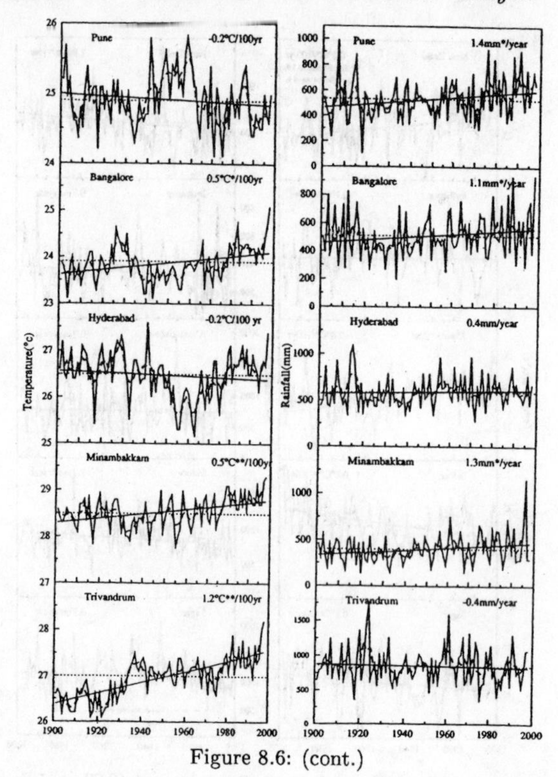

Figure 8.6: (cont.)

warming of over 1.6^oC in winter (DJF) and spring (MAM) seasons (Rupa Kumar and Ashrit, 2001).

In another scenario that includes the effect of sulphate aerosols in addition to the increasing concentration of greenhouse gases, the change in monsoon rainfall is only 7% and the change in the mean annual temperature is 1.1^oC, again with maximum warming of 1.37°C occuring during the winter (DJF) and of 1.14°C occuring during the spring (MAM).

A comparison of several coupled model simulated scenarios for the Indian region is discussed by Giorgi and Francisco (2000).

6. Conclusion

1. Monsoon rainfall in India is trendless—but exhibits strong epochal variations.
2. The observed instrumental records in India show a warming of about $0.5^oC/100$ years for the whole of India.

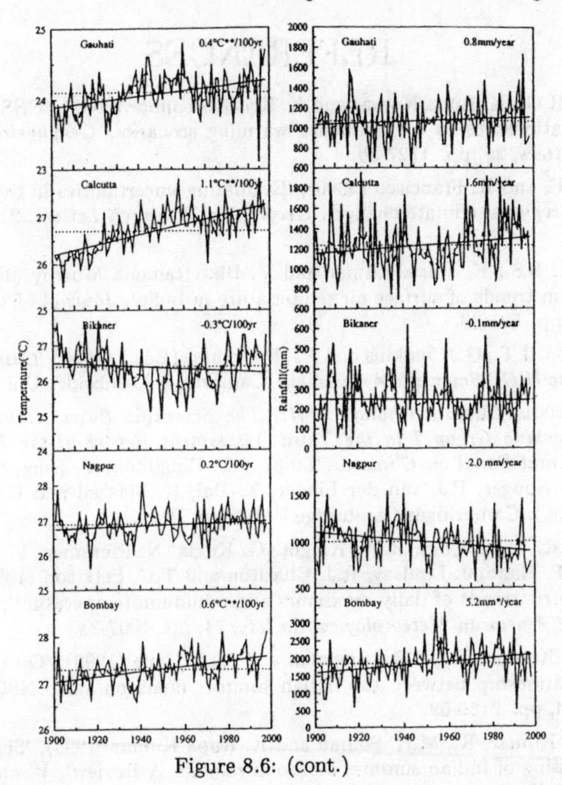

Figure 8.6: (cont.)

3. On interannual time scales the El Niño Southern Oscillation (ENSO) seems to be the most important climate forcing on the monsoon rainfall variability in India.

4. GCM scenarios: Increasing green house gas (GHG) forcing show surface air-temperatures over India to increase by 1.6^oC and the monsoon rainfall by 13%.

5. The scenarios with GHG and sulphate aerosols show subdued trends.

6. In addition to an increase in the mean state of monsoon rainfall, a significant increase in the interannual variability is also noted in the simulations.

7. While there is considerable agreement among models regarding the temperature trends over India, the monsoon rainfall trends differ from model to model.

REFERENCES

Ashrit, R.G., K. Rupa Kumar and K. Krishna Kumar (2001), 'ENSO-Monsoon relationships in a greenhouse warming scenario', *Geophysical Research Letters*, 29, pp. 1727-30.

Giorgi, F. and R. Francisco (2000), 'Evaluating uncertainties in the prediction of regional climate change', *Geophysical Research Letters*, 27, pp. 1295-98.

Hingane, L.S., K. Rupa Kumar and V. Bh. Ramana Murthy (1985), 'Long term trends of surface air temperature in India', *Journal of Climatology*, 5, pp. 521-28.

Houghton, J.T., G.J. Jenkins and J.J. Ephraums (Eds.) (1990), *Climate Change: The IPCC Scientific Assessment*, Cambridge: Cambridge University Press.

IPCC, 2001, *Climate Change 2001: The Scientific Basis, Contribution of Working Group I to the Third Assessment Report of the Intergovernmental Panel on Climate Change*, J.T. Houghton, Y. Ding, D.J. Griggs, M. Nouger, P.J. van der Linder, X. Dai, K. Maskell and C.A. Johnson (eds.), Cambridge: Cambridge University Press.

Karl, T.R., P.D. Jones, R.W. Knight, G. Kukla, N. Plummer, V. Razuvayev, K.P. Gallo, J. Lindsay, R.J. Charlton and T.C. Peterson (1993), 'Asymmetric trends of daily maximum and minimum temperature', *Bulletin of the American Meteorological Society*, 74, pp. 1007-23.

Krishna Kumar, K., B. Rajagopalan and M.A. Cane (1999), 'On the weaking relationship between the Indian summer monsoon and ENSO', *Science*, 284, pp. 2156-59.

Krishna Kumar, K., M.K. Soman and K. Rupa Kumar (1995), 'Seasonal forecasting of Indian summer monsoon rainfall: A Review', *Weather*, 50, pp. 449-67.

Mooley, D.A. and B. Parthasarthy (1984a), 'Fluctuations in all-India Summer Monsoon Rainfall during 1871-1978', *Climatic Change*, 6, pp. 287-301.

Mooley, D.A. and B. Parthasarthy (1984b), 'Variability of the Indian Summer Monsoon and Tropical Circulation features', *Monthly Weather Review*, 111, pp. 967-78.

Pant, G.B. and L.S. Hingane (1988), 'Climatic Changes in and around the Rajasthan Desert during 20th century', *Journal of Climatology*, 8, pp. 391-401.

Pant, G.B. and K. Rupa Kumar (1997), *Climates of South Asia*, London: John Wiley & Sons.

Pant, G.B., H.N. Bhalme, B. Parthasarthy, L.S. Hingane, and K. Rupa Kumar (1991), 'Changes in Climate over India and adjacent areas since the last ice age', Chapter XVIII of IPCC (WMO/UNEP) Scientific Assessment Report Supplement.

Pant, G.B., K. Rupa Kumar and B. Parthasarthy (1993), Observed Variations in Rainfall and Surface Temperature over India, in M. Lal (ed.), *Global Warming, Concern for Tomorrow*, New Delhi: Tata-McGraw Hill.

Pant, G.B., K. Rupa Kumar, B. Parthasarthy and H.P. Borgaonkar (1988), 'Long term variability of the Indian Summer Monsoon and related parameters', *Advances in Atmospheric Science*, 1, pp. 469-485.

Parthasarthy, B. and O.N. Dhar (1978), 'Climatic fluctuations over Indian Region Rainfall : A Review', Research report No. RR-025, Indian Institute of Tropical Meteorology, Pune, India.

Parthasarthy, B., K. Rupa Kumar and A.A. Munot (1993), 'Homogeneous Indian Monsoon Rainfall: Variability and Prediction', *Procedings of the Indian Academy of Science (Earth and Planetary Sciences)*, 102, pp. 121-55.

Parthasarthy, B., N.A. Sontakke, A.A. Munot and D.R. Kothawale (1987), 'Drought/Floods in Summer Monsoon Season over different Meteorological subdivisions of India for the period 1871-1984', *Journal of Climatology*, 7, pp. 57-70.

Rupa Kumar, K. and R.G. Ashrit (2001), 'Regional Aspects of Global Climate Change Simulations: Validation and Assessment of Climate Response of the Indian Monsoon Region to Transient Increase of Green House Gases and Sulphate Aerosols', *Mausam*, 52, pp. 229-44.

Rupa Kumar K. and L.S. Hingane (1988), 'Long-term Variations of Surface Air Temperature at Major Industrial Cities of India, *Climatic Change*, 13, pp. 287-307.

Rupa Kumar K., K. Krishna Kumar and G.B. Pant (1994), 'Diurnal Asymmetry of Surface Temperature Trends over India', *Geophysical Research Letters*, 21, pp. 677-80.

Rupa Kumar K., Pant, G.B., Parthasarthy, B. and Sontakke, N.A. (1992), Spatial and sub-seasonal patterns of the long term trends of Indian summer monsoon rainfall, *International Journal of Climatology*, 12, pp. 257-68.

Sarker, R.P. and Thapliyal, V. (1988), 'Climate Change and Variability', *Mausam*, 39, pp. 127-38.

Soman, M.K., K. Krishna Kumar and N. Singh (1988), 'Decreasing Trend in the Rainfall of Kerala', *Current Science*, 57, pp. 7-12.

Srivastava, H.N., B.N. Dewan, S.K. Dikxhit, G.S. Prakash Rao, S.S. Singh and K.R. Rao (1992), 'Decadal Trends in Climate over India', *Mausam*, 43, pp. 7-20.

Thapliyal, V. and S.M. Kulshrestha (1991), 'Climate Changes and Trends over India', *Mausam*, 42, pp. 333-58.

Walker, G.T. (1924), 'Correlation in Seasonal Variations of Weather - IX : A Further Study of World Weather (World Weather II)', *Memoirs of India Meteorological Department*, New Delhi, 24, pp. 275-332.

9

Data Assimilation, an Important and Challenging Research Topic for Environmental Statisticians

XIAOGU ZHENG

Data assimilation, which aims at deriving good initial states by using observations and governing equations of physical systems, plays a key role in the numerical prediction of environmental states. The major challenges in data assimilation for environmental statisticians are:

(1) how to apply the prior knowledge of model and observations to derive idealized equations for best initial states; and

(2) how to approximate the idealized equations for avoiding processing large-dimensional matrices.

This paper provides a brief introduction to data assimilation for environmental statisticians. Some views on collaboration between environmental statisticians and environmental scientists are also expressed.

This study was supported by a Visiting Fellowship of the Institute of Statistical Mathematics offered by the Japanese Ministry of Education, Science, Sports and Culture. The invitation from T. Ozaki is greatly appreciated. I wish to thank J. A. Renwick for his valuable comments.

1. Introduction

Variations in the atmosphere and ocean on time and space scales are important to humans, and pretty much define the 'environment' along with the biosphere. The prediction of meteorological and oceanic states is an important research field in environmental sciences. This is partly because forecasting is one of the main thrusts of earth science in general, and partly because good forecasts are of huge practical benefit. Data assimilation plays a key role in numerical prediction.

It is well known that the earth system is chaotic, that is, it is governed by non-linear partial differential equations (PDEs), but the solutions are very sensitive to initial conditions (Lorenz 1963). Observations are not good initial states, because not all model variables are observed and there exist observation errors. Data assimilation is then proposed for estimating the true states of the earth system by fitting observations to the governing PDEs. Better predictions can be derived starting from the estimated true initial states. Increasingly, it is being recognized through the constant confrontation of theory (in the form of more general earth system models) with reality (as provided by earth system data) represented by the data assimilation process, that major advances can be expected in the scientific understanding of the dynamics, variability, and interactions of all components of Earth Systems over a broad range of time and space scales.

The desire to carry out data assimilation is also being driven largely by the rapidly increasing amount of new forms of observational data (especially, data observed by the space-borne platforms) becoming available, by pressing scientific and social needs to understand the behaviour of the earth system as a whole. At the same time, the ability to pursue data assimilation in a physically and mathematically sound fashion is being enabled by the increasing sophistication of earth system models and by rapid advances in computing technology.

There are two important components of data assimilation: numerical modelling and statistical optimization. Numerical modelling includes building forecast models with fewer system errors and determining the initial states for suppressing high frequency inertia-gravity wave motions (Daley 1991). Numerical modelling is related mainly to dynamics; statisticians have little to do on it. Statistical optimization tries to characterize model errors (which is defined as the difference between the discrete dynamics and the governing continuum dynamics and is represented by statistical forcing) and the observational errors, to avoid processing matrixes with large dimensions, to tune the free parameters, thereby offering a rigorous, data-driven means of improving our earth system modelling capability. Statisticians have a large role to play regarding these problems.

Most meteorologists publish their research results in meteorological journals. Since these journals are inconvenient for statisticians to access, we recommend the monograph 'Atmospheric data analysis' (Daley 1991) for reference regarding basic knowledge, and Vol. 75, No. 1B, *Journal of the Meteorological Society of Japan* (1997) for reference with regard to more advanced developments of data assimilation.

Statisticians want to work with geophysicists (here I mean meteorologists and oceanographers). Since the research attitudes and backgrounds of geophysicists (I mean meteorologists and oceanographers) and statisticians are different, there is a need to bridge the gap between these disciplines. Work on data assimilation is a research topic that involves the application of both disciplines and can help communication between the two fields.

To the best of my knowledge, Professor Wahba (University of Wisconsin, Madison) and Professor Thiebaux (Dalhousie University, Halifax) are the two leading statisticians for data assimilation. Wahba and her students focus on applying generalized cross-validation to variational assimilation (Wahba *et al.* 1995). Thiebaux has been involving in development of an operational assimilation scheme (Thiebaux *et al.* 1986). Their contributions are well respected by geophysicists.

This paper is arranged as follows. The concept of data assimilation is introduced in Section 2. Section 3 is devoted to data assimilation methods. Our conclusion is presented in Section 4.

2. Basic Concepts and Definitions

A non-linear dynamical system can be expressed by the ordinary differential equation:

$$\frac{d}{dt}x(t) = f(x(t)) \tag{9.1}$$

where $x(t)$ is the state vector at time t and f is the non-linear dynamical function. The goal of data assimilation is to find a series of analysis states which are sufficiently close to the true states by using the information provided by the dynamical system and the observations.

Following the notation of Ide *et al.* (1997), the superscripts t, o, a and f denote 'truth', 'observation', 'analysis' and 'forecast' respectively. The discrete governing equation that is associated with Eq. (9.1) is

$$x^t(t_{i+1}) = M_i[x^t(t_i)] + \eta(t_i) \tag{9.2}$$

where $x^t(t_{i+1})$ is the true state, M_i is a multivariate function and $\eta(t_i)$ is the system error vector which is normally distributed with mean zero and system error covariance matrix $Q(t_i)$. The observational equation is

$$y_i^o = H_i x^t(t_i) + \varepsilon_i \tag{9.3}$$

where y^{0_i} is the observational vector, H_i is the observational matrix and ε_i is the observational error which is normally distributed with mean zero and observational covariance matrix R_i. Note that both $x^t(t_{i+1})$ and y_i^o are random because $\varepsilon(t_i)$ and a_i^o are random.

In principle, an analysis state $x^a(t_i)$ is defined by the conditional mean of the true state $x^t(t_i)$, given observations $y_0^o, \ldots, y_i^o$ (Jazwiski 1970). For a given analysis state at time t_{i-1}, the forecast from time t_{i-1} to time t_i is

$$x^f(t_i) = M_{i-1}[x^a(t_{i-1})] \tag{9.4}$$

with forecast error covariance matrix $P^f(t_i)$. By the theory of extended Kalman filtering (EKF) (Jazwinski 1970), the conditional mean of $x^t(t_i)$, given $y_0^o, \ldots, y_i^o$, is

$$x^a(t_i) = x^f(t_i) - K_i(H_i x^f(t_i) - y_i^o) \tag{9.5}$$

where

$$K_i = P^f(t_i)H_i^T(H_i P^f(t_i) H_i^T + R_i)^{-1} \tag{9.6}$$

is the Kalman gain matrix. The analysis error covariance matrix is

$$P^a(t_i) = (I - K_i H_i)P^f(t_i) \tag{9.7}$$

There are two major obstacles for the implementation of Eqs.(9.5) −(9.6). First, the estimation of the forecast error covariance matrix $P^f(t_i)$ is not an easy task. Second, processing a large-dimensional matrix has to be avoided, because the dimensions of model states and observations can be several millions or more.

3. Assimilation Schemes

In this section, we introduce four popular assimilation schemes: optimal interpolation (OI), three-dimensional variational method (3D-Var), four-dimensional variational method (4D-Var) and extended Kalman filtering (EKF) (Daley 1991). The methods OI and 3D-Var are more operational in nature than the others. The strategies they use to overcome the two major obstacles are different. The methods 4D-Var and EKF are more advanced theoretically, but their implementations are more difficult.

3.1. Optimal Interpolation

The method OI is a sequential one (Daley 1991). It applies regionally local statistical interpolation to overcome the two obstacles. It was developed rapidly during the 1980s. It was, until recently, the most widespread scheme in operational use for weather-prediction and oceanic data assimilation.

In the OI scheme, the forecast error covariance matrix $P^f(t_i)$ is replaced by an approximation $B^f(t_i)$ which can be expressed as a product of variances, placed in a diagonal matrix $D^f(t_i)$, and of correlations, placed in a matrix C with unit diagonal

$$B^f(t_i) = [D^f(t_i)]^{1/2}C[D^f(t_i)]^{1/2} \tag{9.8}$$

A simplified prediction scheme for the error variances alone is introduced, while C is kept constant in time:

$$D^f(t_i) = N[D^a(t_{i-1})] \tag{9.9}$$

Various simple schemes N have been implemented to compute D^f.

The Kalman gain matrix K_i is approximated by

$$K_i^{OI} = B^f(t_i)H_i^T[H_iB^f(t_i)H_i^T + R_i]^{-1} \tag{9.10}$$

The analysis state is estimated as

$$x^a(t_i) = x^f(t_i) - K_i^{OI}(H_i x^f(t_i) - y_i^o) \tag{9.11}$$

The analysis error variance for $D^a(t_i)$ is estimated by

$$D^a(t_i) = \text{diag}[B^a(t_i)] \tag{9.12}$$

where

$$\begin{aligned} B^a(t_i) &= (I - K^{OI}H_i)B^f(t_i)(I - K^{OI}H_i)^T \\ &\quad + K^{OI}R_i(K^{OI})^T \end{aligned} \tag{9.13}$$

Two salient features of OI are:

(a) A local approximation in Eq. (9.13), either by selecting only a few observations near each grid point that has been updated or by decomposing the domain into regular, small subdomains and reflecting the influence of the observations outside each subdomain.

(b) A resulting approximation in Eqs. (9.10) and (9.13) by using direct evaluation of a correlation model C at the selected observation locations, to yield $H_i B^f(t_i) H_i^T$.

Although OI is computationally economical, the derived analysis errors may be generally large. This is mainly because the approximations introduced are not good enough. During the 1990s, great efforts were made to develop alternative schemes, mainly variational methods.

3.2. Three-Dimensional Variational Method

The variational formalism provides an elegant framework for data assimilation. 3D-Var is a variational method. It was not considered to be very useful in practice before the 1990s, because of the computationally intensive nature of the resulting algorithms. It was rapidly developed during that decade due to the greatly increased computing capacity that became available. 3D-Var is regarded as a more advanced data assimilation scheme than OI. Different 3D-Var varsions were implemented in major meteorological centres, such as European Center for Medium Range Weather Forecasting, US National Center for Environmental Prediction and UK Meteorological Office.

The theoretical base of 3D-Var is that the conditional mean of $x^t(t_i)$ given $y_0^o, \ldots, y_i^o$ [that is Eqs. (9.5) and (9.6)] minimizes the cost function

$$\begin{aligned} J_i(x^t(t_i)) &= \frac{1}{2}[x^t(t_i) - x^f(t_i)]^T[P^f(t_i)]^{-1} \\ & [x^t(t_i) - x^f(t_i)] + \frac{1}{2}(H_i x^t(t_i) - y_i^o)^T \\ & R_i^{-1}(H_i x^t(t_i) - y_i^o) \end{aligned} \tag{9.14}$$

Given x^f and $(P^f)^{-1}$, the cost function $J_i(x^t(t_i))$ is a quadratic function for x^t. Then, the minimum point can be estimated by the localized quasi-Newton method (Daley 1991). In this way, processing large-dimensional matrices are avoided. The key point for the Newton method is to derive Jacobian and Hensian of the cost function. In 3D-Var, x^f is approximated by the background state x^b which is an operational prediction.

A major difficulty for 3D-Var is to estimate $(P^f)^{-1}$. It is well-known that if $(P^f)^{-1}$ is not estimated appropriately, the estimated analysis states can be very wrong. In current 3D-Var schemes, $(P^f)^{-1}$ is estimated empirically. To improve the estimation will be a major objective for 3D-Var in the future.

3.3. Four-Dimensional Variational Method

The four-dimentional variational method (4D-Var) demands a close fit to the data, plus consistency with a dynamic model over an extended period of time. Therefore, it is regarded as a more advanced assimilation scheme than 3D-Var. It was first implemented in European Center for Medium Range Weather Forecasting. Recently it was also implemented successfully in UK Meteorological Office.

The cost function for 4D-Var is (Daley 1991)

$$\begin{aligned} J(x^t(t_i)) &= \frac{1}{2}(x^t(t_i) - x^f(t_i))^T[P(t_i)]^{-1}(x^t(t_i) \\ & -x^f(t_i)) + \frac{1}{2}\sum_{j=0}^{n}\{(H_{i+j}[M(t_{i+j}, t_i)x_i^t] \\ & -y_{i+j}^o)^T R_{i+j}^{-1}(H_{i+j}[M(t_{i+j}, t_i)x_i^t] \\ & -y_{i+j}^o)\} \end{aligned} \tag{9.15}$$

where n is the step for the extended time period

$$M(t_i, t_i) = I, M(t_{i+j}, t_i) = \prod_{k=i}^{i+j-1} M(t_{k+1}, t_k) \tag{9.16}$$

and $M(t_{i+1}, t_i)$ is a linearized operator of dynamics to advance a state from t_i to t_{i+1}, that is, a matrix M_i which depends on x such that

$$M(t_{i+1}, t_i)(x) = M_i x \tag{9.17}$$

When $n = 0$, Eq. (9.15) reduces to Eq. (9.14) and 3D-Var is therefore a special case of 4D-Var. Under the assumption of no system error, that is, η in Eq. (9.2) is zero, $M(t_{i+j}, t_i)[x^t(t_i)]$ would be close to the true state $x^t(t_{i+j})$ and the term

$$\begin{aligned} \frac{1}{2}\sum_{j=1}^{n} \{ & (H_{i+j}[M(t_{i+j}, t_i)x_i^t] - y_{i+j}^o)^T R_{i+j}^{-1} \\ & (H_{i+j}[M(t_{i+j}, t_i)x_i^t] - y_{i+j}^o)\} \end{aligned} \tag{9.18}$$

adds consistency with a dynamic model over an extended period of time $[t_{i+1}, t_{i+n}]$ to the 3D-Var cost function. Therefore, analysis states derived by 4D-Var should be better than 3D-Var.

However, if system error η is significant, $M(t_{i+j}, t_i)[x^t(t_i)]$ would not be close to $x^t(t_{i+j})$. Then Eq. (9.18) will not represent any dynamical constraint. In this case, analysis states dervied by 4D-Var methods may be worse than those derived by 3D-Var. Meteorologists have to suppose the 'no system error' assumption to be roughly true for some numerical forecasting models. But up to now, analysis states derived by 4D-Var have not been significantly better than those derived by 3D-Var. Another difficulty is that the implementation for minimizing 4D-Var cost function is more difficult than that for the 3D-Var cost function. In fact, the adjoint technique is applied (Daley 1991).

3.4. Extended Kalman Filtering

The major advantage of the Extended Kalman Filtering (EKF) method is that it can estimate forecast error covariance matrix if the system error covariance matrix Q is known. For EKF assimilation, the forecast error covariance matrix P^f can be estimated by solving the equation

$$\frac{dP^f(t)}{dt} = F(t)P^f(t) + P^f(t)F(t) + Q(t) \tag{9.19}$$

where the function $F(t)$ is the Jacobian of the dynamics f at the initial state defined by $x^t(t)$. Some approximate solution of Eq. (9.19) may be derived under some simplified constraints (Ide and Ghil 1997).

If there is no system error, EKF is equivalent to 4D-Var. If systems are small and parameters are known exactly, system error covariance matrices can be derived by simulations. However, earth systems are large and their parameters are subject to errors. System error covariance matrices for the earth system are not easy to estimate. This is the major obstacle to applying EKF to meteorological data assimilation. However, EKF may be applicable to data assimilation with simplified physical models, such as the Zebiak-Cane model for predicting the El Niño southern oscillation (Zebiak and Cane 1987), where the system error covariance matrix is derived using wind stress data.

4. Conclusions

This paper has introuduced the basic concept of data assimilation for earth systems. The current developments, the major difficulties and the possible future work are also discussed. Because data assimilation is a key step for the numerical prediction of environmental states and it combines large-dimensional non-linear dynamics and non-linear statistical techniques, it is one of the most important and challenging research fields in environmental sciences. The environmental statistician will have a big role to play in data assimilation. Any essential progress regarding this will be greatly appreciated by environmental scientists.

REFERENCES

Daley, R. (1991), *Atmospheric Data Analysis,* Cambridge: Cambridge Univerisity Press.

Ide, K., P. Courtier, M. Ghil and A. C. Lorenc, (1997), 'Unified Nation for Data Assimilation: Operational, Sequential and Variational', *Journal of the Meteorological Society of Japan*, 75, pp. 181-9.

Ide, K. and M. Ghil (1997), 'Extended Kalman Filtering for Vortex Systems. Part I: Methodology and Point Vortices', *Dynamics of Atmospheres and Oceans*, 27, pp. 301-32.

Jazwiski, A. H. (1970), *Stochastic Processes and Filtering Theory,* London: Academic Press.

Lorenz, E. N. (1963), 'Deterministic Non-periodic Flow', *The Journal of Atmospheric Sciences*, 20, pp. 130-41.

Thiebaux, H. J., H. Mitchell and D. Shantz (1986), 'Horizontal Structure of Hemispheric Forecast Error Correlations for Geopotential and Temperature', *Monthly Weather Review*, 114, pp. 1048-66.

Wahba G., D. R. Johnson, F. Gao and J. Gong (1995), 'Adaptive Tuning of Numerical Weather Prediction Models : Randomized GCV in Three and Four dimensional Data Assimilation', *Monthly Weather Review*, 123, pp. 3358-69.

Zebiak, S. E. and M. A. Cane (1987), 'A Model El Niño-Southern Oscillation', *Monthly Weather Review*, 115, pp. 2262-78.